Springer Proceedings in Mathematics & Statistics

Volume 372

This book series features volumes composed of selected contributions from workshops and conferences in all areas of current research in mathematics and statistics, including operation research and optimization. In addition to an overall evaluation of the interest, scientific quality, and timeliness of each proposal at the hands of the publisher, individual contributions are all refereed to the high quality standards of leading journals in the field. Thus, this series provides the research community with well-edited, authoritative reports on developments in the most exciting areas of mathematical and statistical research today.

More information about this series at http://www.springer.com/series/10533

Abdeljalil Nachaoui · Abdelilah Hakim ·
Amine Laghrib
Editors

Mathematical Control and Numerical Applications

JANO13, Khouribga, Morocco, February 22–24, 2021

Springer

Editors
Abdeljalil Nachaoui
Laboratoire de Mathématiques Jean Leray
Université de Nantes
Nantes, France

Abdelilah Hakim
Département Mathématiques
Université Cadi Ayyad
Marrakech, Morocco

Amine Laghrib
FST Béni Mellal
Université Sultan Moulay Slimane
Béni Mellal, Morocco

ISSN 2194-1009 ISSN 2194-1017 (electronic)
Springer Proceedings in Mathematics & Statistics
ISBN 978-3-030-83441-8 ISBN 978-3-030-83442-5 (eBook)
https://doi.org/10.1007/978-3-030-83442-5

Mathematics Subject Classification: 49M37, 68Q99, 49-06

This Springer imprint is published by the registered company Springer Nature Switzerland AG
The registered company address is: Gewerbestrasse 11, 6330 Cham, Switzerland

Preface

This book contains a collection of selected papers presented at the *Numerical Analysis and Optimization Days* (JANO13), held in Khouribga, Morocco, from 22 to 24 February 2021. The book focuses on topics related to mathematical control and numerical applications. This volume covers new trends and advances in several very important fields of mathematical modelling and recent developments in new numerical approximation methods. The relevance of these topics is particularly related to the study of approximation parameters, sources separation and partial differential equations (PDEs) approximations related to image processing applications. All contributing authors are eminent researchers and scholars in their respective fields, hailing from around the world. The information given in this book will be of value to the work of researchers, scientists and engineers in the fields of both academia and industry.

Nantes, France — Abdeljalil Nachaoui
Marrakech, Morocco — Abdelilah Hakim
Béni Mellal, Morocco — Amine Laghrib

Contents

VRP with Flexible Time Windows Using the ALNS Metaheuristic Algorithm

Fatah Labdiad, Mehdi Nasri, Hamza Khalfi, and Imad Hafidi

Abstract Vehicle routing problems are subject to massive investigations in operations research. We address in this presentation a specific variant of the problem called in the literature vehicle routing problem with flexible time windows (VRPFlexTW), when the solution has to obey several other constraints, such as the consideration of travel, service, and waiting times together with time-window restrictions. An interesting wildly used algorithm is the Adaptive Large Neighbourhood Search (ALNS), we describe the ALNS approach and its different components and we list a computational comparison between the ALNS and the Ant colony optimiser (ACO) on few instances of the VRPFlexTW.

Keywords Vehicle routing problem · Flexible time windows · ALNS algorithm · Meta-heuristics · Optimization

The Vehicle Routing Problem (VRP) is described as the problem of finding an optimal collection of routes from one or several depots to a predetermined number of scattered locations subject to side constraints enforcing some given importance criteria relative to cost, time, distance or a combination of these quantities. The basic version of the VRP problem is an extension of the Travelling Salesman problem [1]. It was originally introduced by [2] under the name of "Truck Dispatching Problem" and since then carried out the object of many intensive studies in its modelling and resolution aspects. VRPs nowadays plays a central role in many fields and in some real world application among which we cite the physical distribution and logistics, supply chain management, finance and so on. There exists a wide variety of VRPs and a broad literature on this class of problems see for example [3–6]. At its basic form, a VRP can be viewed as: a fleet of vehicles located at a central depot, that must ensure tours between several customers who have requested a certain merchandise or service. The set of customers visited by a vehicle refers to its tour and each tour

F. Labdiad · M. Nasri · H. Khalfi (✉) · I. Hafidi
LIPIM, ENSA Khouribga, Université Sultan Moulay Slimane, Khouribga, Morocco
e-mail: h.khalfi@usms.ma

I. Hafidi
e-mail: i.hafidi@usms.ma

A. Nachaoui et al. (eds.), *Mathematical Control and Numerical Applications*,
Springer Proceedings in Mathematics & Statistics 372,
https://doi.org/10.1007/978-3-030-83442-5_1

starts and ends at the central depot. Each customer must be served once and only once and by one and only one vehicle. The objective of the standard VRP model is to minimize the sum of the distances travelled or the total travel time of vehicle rounds while meeting customer demand.

The model can be represented as a closed graph $G = (V, A)$ [7] where vertexes are clients V = 0,1,....,n and 0 denotes the origin depot, arcs are the routes i, j linking two clients. There are m binary variables x_{ijp} used to check if a trajectory (ij) is actually travelled by the vehicle p or not. A second binary variable y_{ip} is to enforce the condition that each client will be served by only one vehicle, hence y_{ip} is equal to 1 when the vehicle p visited the node i and 0 otherwise. The mathematical model could be formulated as follows:

$$\begin{cases} Z = \min / \max F \quad (1) \\ \text{s.c.} \sum_{p=1}^{m} \sum_{i=1}^{n-1} x_{0ip} \leq m \quad (2) \\ \sum_{p=1}^{m} \sum_{i=1}^{n-1} x_{i0p} \leq m \quad (3) \\ \sum_{p=1}^{m} y_{kp} \leq 1, \ \forall k = 1, \ldots, n \quad (4) \\ \sum_{j=1}^{n-1} x_{ijp} = y_{ip}, \ i = \{1, \ldots, n\} \ p = \{1, .., m\} \quad (5) \\ \sum_{j=1}^{n} x_{jip} = y_{ip}, \ i = \{1, \ldots, n\} \ p = \{1, .., m\} \quad (6) \\ x_{ijp}, y_{ip} \in \{0, 1\}, \ \forall i, j = \{1, \ldots, n\}, p = \{1, \ldots, m\} \quad (7) \end{cases} \tag{1}$$

The constraints (2) and (3) ensure that the number of vehicles leaving the depot is the same as the number of vehicles entering the depot. The constraint (4) ensures that each client from 1 to n is visited by a maximum of one vehicle. The constraints (5) and (6) represent the conservation of flow for each client i and ensure that the number of vehicles crossing all the arcs entering $\{(j, i), \forall j \in A\}$ is equal to the number of vehicles crossing the outgoing arcs $\{(i, j), \forall j \in A\}$. Finally, the binary variables used x_{ijp} and y_{ip} are declared by the constraint (7).

The vehicle routing problem is a classic extension of the travelling salesman problem. Both are part of the class of NP-complete problems. It is part of the optimization problems for which we do not know an algorithm allowing to find an exact solution quickly (polynomial time) in all cases.

As the VRP appears in real life, it may have several classes of additional constraints, such as limits on the vehicle capacity [8, 9], time windows for serving customers [3–5], route lengths, or the number of hours worked by a driver or a distribution clerk. For a recent complete review on the classification of different VRP variants, see [6, 10]. As with basic VRP, most VRP variants are known to be NP-hard.

In this paper, we are interested in the VRP with flexible time window (VRPFlexTW). It is a relaxation of the VRPTW where time windows are considered hard constraints that should not be violated.

The layout of the paper is as follows: the next section aim to provide a brief introduction to the multi-objective VRPFlexTW problem modelling and a comprehensive overview of the popular resolution techniques used in the literature. Section 2 will discuss the ALNS algorithm and its components with application to the considered VRPFLexTW problem. In Sect. 3, numerical results will be presented and a comparison with other techniques. Eventually, a summary of the results and discussion will be provided to wrap up this study.

1 Problem Formulation of the VRPFlexTW

Let us consider a multi-objective VRPFlexTW formulation that seeks to optimize customer satisfaction when vehicle routes are constrained by capacity and time windows, while minimizing costs associated with the distance travelled and the number of vehicles. When one wants to extend the previous model (1) to the flexible version of the VRPFlexTW, it is necessary to make some modifications to the mathematical formulation. First of all, remember that a time window is in fact a time interval in which it is allowed to serve a customer at no additional cost. In our case, this interval is flexible, that is, with a certain penalty, it is possible to perform the service to customers outside this interval. It therefore becomes possible to increase the time windows for meeting clients from $[a_i, b_i]\forall i \in N$ to $\left[a_i - a_i', b_i + b_i'\right], i \in N$. The constants a_i' and b_i' satisfy $a_i - a_i' \geq E_i$ and $b_i + b_i' \leq L_i$ where E_i and L_i are respectively tolerances for serving clients earlier or later than the appointed time interval. Although the waiting time is permitted at no cost, a client's satisfaction denoted by $\mu_i(z_i)$ will be constant on the interval $[a_i, b_i]$ but will decrease to 0 linearly when the time moves away from the agreed upon interval limits. The satisfaction function μ_i is taken as follows :

$$\mu_i(z_i) = \begin{cases} 0, & z_i < E_i \\ \frac{z_i - E_i}{a_i - E_i}, & E_i \leq z_i < a_i \\ 1, & a_i \leq z_i \leq b_i \\ \frac{L_i - z_i}{L_i - b_i}, & b_i < z_i \leq L_i \\ 0, & z_i > L_i \end{cases} \tag{2}$$

Before presenting the mathematical formulation, we define the following notations:

- h_k is the transportation cost per unit distance of vehicle k
- f_k is the fixed cost incurred for using vehicle k
- c_{ij} is the distance between vertex i and vertex j,
- s_i is the service time at vertex i

- w_i is the waiting time at vertex i
- t_{ij} is the time required for travelling from vertex i to vertex j
- Decision variables:

$$x_{ijk} = \begin{cases} 1, & \text{if vehicle } k \text{ travels from vertex } i \text{ to vertex } j \\ 0, & \text{otherwise} \end{cases} \tag{3}$$

$$y_{ik} = \begin{cases} 1, & \text{if vertex } i \text{ is served by vehicle } k \\ 0, & \text{otherwise} \end{cases} \tag{4}$$

Given the above parameters and decision variables, the problem can be formulated as follows:

$$\max \frac{1}{n} \sum_{i=1}^{n} \mu_i (z_i) \tag{5}$$

$$\min \sum_{k=1}^{m} h_k \sum_{i=0}^{n} \sum_{j=0}^{n} c_{ij} x_{ijk} + \sum_{k=1}^{m} f_k \sum_{j=1}^{n} x_{0jk} \tag{6}$$

$$\sum_{i=0}^{n} x_{ijk} = y_{jk} \quad \forall k \in \{1, 2, \cdots, m\}, \quad \forall j \in \{1, 2, \cdots, n\} \tag{7}$$

$$\sum_{j=0}^{n} x_{ijk} = y_{ik} \quad \forall k \in \{1, 2, \cdots, m\}, \quad \forall i \in \{1, 2, \cdots, n\} \tag{8}$$

$$\sum_{i=0}^{n} \sum_{j=0}^{n} x_{ijk} \left(t_{ij} + s_i + w_i \right) \le r_k \quad \forall k \in \{1, 2, \cdots, m\} \tag{9}$$

$$w_0 = s_0 = 0 \tag{10}$$

$$\sum_{k=1}^{m} \sum_{i=0}^{n} x_{ijk} \left(z_i + w_i + s_i + t_{ij} \right) = z_j \quad \forall j \in \{1, 2, \cdots, n\} \tag{11}$$

$$E_i \le z_i + w_i \le L_i \quad \forall i \in \{1, 2, \cdots, n\} \tag{12}$$

$$w_i = \max \{0, E_i - z_i\} \quad \forall i \in \{1, 2, \cdots, n\} \tag{13}$$

$$x_{ijk} \in \{0, 1\} \quad \forall i, j \in \{1, 2, \cdots, n\}, \quad \forall k \in \{1, 2, \cdots, m\} \tag{14}$$

$$y_{ik} \in \{0, 1\} \quad \forall i \in \{1, 2, \cdots, n\}, \quad \forall k \in \{1, 2, \cdots, m\} \tag{15}$$

$$z_i \ge 0 \quad \forall i \in \{1, 2, \cdots, n\} \tag{16}$$

In the model above, Objective (2) is to maximize the customer satisfaction. Objective (3) is to minimize the total routing costs, which consist of travel costs and fixed vehicle costs. Constraint (4) guarantees that the vehicle capacity is not exceeded; Constraint (5) ensures that each customer is served by exactly one vehicle; and Constraint (6) ensures that each route starts and ends at the depot. Constraints (7)-(8) guarantee that each customer is served exactly once. Constraint (9) ensures that the maximum route time is not exceeded; Constraint (10) defines the waiting and service time at the depot; Constraint (11) represents the relationship between the arrival time at a vertex and the departure time from its predecessor; Constraint (12) ensures that customers are served within the required time; and Constraint(13) defines the waiting time.

2 ALNS State of the Art in the Context of VRP

Since the publication of the seminal work of [2], the Vehicle Routing Problem (VRP) has been the subject of a great deal of research. The VRP has extended its scope by introducing additional real-world facets to derive a large number of variants different from the one introduced by Dantzig and generalized by [11]. For given customers with known locations and demands, the goal of the VRP problem in its general form is to design a set of minimum cost routes that serve a number of geographically distributed customers around the central depot, and fulfilling specific constraints of the problem.

The VRP belongs to the class of operations research and combinatorial optimization problems. Exact method fails to solve this problem of optimization for large instances more than one hundred customers except the Traveling Salesman Problem (TSP) where instances with several thousand nodes can be solved optimally on a regular basis, see for instance [12, 13]. The VRP is an NP-hard problem, see for e.g. [14]. It is hard to solve problem with small instances optimally above 50 customers as reported by [15].

For a comprehensive state of the art classification and review of the VRP, we refer the reader to the paper of [16] and to an adapted version of the taxonomy suggested by [17]. Another recent concise review of existing VRP problem features and applications can be found in the paper by [18].

The famous extension of the VRP is the VRP with Time Windows (VRPTW) ([19–21]), where the service at any customer starts within a certain time interval called a time window. Time windows are defined as hard (or strict) when it must be satisfied, while it is called soft if it can be violated and allow deliveries outside the boundaries against a penalty added to the objective function.

Combinatorial optimization problems of this kind are non-polynomial-hard (NP-hard) and are hence best solved by using heuristics and metaheuristics which are often more suitable for real-life problems which are large scale in nature. Their complexity may emanate from various sources, such as customers, vehicles, shipments and physical infrastructure, all interacting in different ways. The most important metaheuristics used to solve the VRPTW are Tabu Search (TS) ([22–24]), the Simulated Annealing (SA) ([25, 26]), the Adaptive Large Neighborhood Search (ALNS)

([27]), Genetic Algorithm (GA) ([28, 29]), Ant Colony Optimisation Algorithm (ACO) ([30, 31]) and Variable Neighborhood Search (VNS) ([32, 33]). As survey of metaheuristics for VRPTW can be found in [19].

The survey of [20] is given in the context of column generation in general and the focus is therefore the path formulation of the VRPTW. These surveys also give a status on the computational success of the state of the art algorithms proposed in the literature. An other interesting way to treat such problem is to combine metaheuristics, mathematical programming, constraint programming and machine learning, see for e.g. [34].

Over time, researchers have developed more powerful algorithms that find improved solutions to both real-world VRPTW instances and well studied [28, 35] benchmark instances by building on previous work and by taking advantage of faster computational resources.

For real problems, the instances are massive scale and the input data changes in an unexpected way, the major challenge is to optimize the execution time of the algorithm. Therefore, the need of the parallelization has become an indispensable and even inevitable technique to tackle this problem. But, it is important to keep in mind that the relevant noun counterpart which is the quality of the solution can be affected. For instance, we try to select a good solution in a reasonable time without too much compromising the quality of the solution. An efficient method to find a compromise between the time of execution and the quality of a solution will be studied in this work.

Following these ideas, [36] has proposed parallel ALNS applied to the Travel Salesman Problem with Pickup and Delivery, and the Capacitated Vehicle Routing Problem such that each worker thread obtains a copy of the current solution and performs destroy and repair operations on its local copy in order to generate a local solution which will be compared to the global and best found solutions. In the same spirit, [37] has proposed a parallel variant of the Large Neighborhood Search LNS to solve the Multi-depot and Periodic Vehicle Routing Problem. On the same topic, [38] has presented a parallel version of the ALNS by adding a set covering post-optimization model that combines the tours generated throughout the search to assemble a better solution.

We devote the next section to the exposition of the Adaptive Large Neighborhood Search meta heuristic introduced by [39], including the presentation of different destruction and insertion heuristics involved therein.

3 ALNS Techniques for VRPFlexTW

The ALNS is a metaheuristic proposed by [40] as an extension of the Large Neighborhood Search (LNS) heuristic presented by [41]. It is an adaptive approach employed to ameliorate an incumbent solution. The basic idea of this algorithm is to improve a given initial solution by exploring many large neighborhoods. This can be done by the application of various destroy and repair operators. More precisely, the destroy operator removes nodes from the solution. We obtain then an incomplete or infeasible solution d(x). The repair operator reinserts the removed nodes at more favored position which leads to a new feasible (complete) solution r(d(x)). The resulting solution will be accepted or rejected based on a Hill Climbing acceptance criteria which only accepts solutions that are better than the current one. This process will be terminated when a stop criterion is met.

It is worth mentioning, that ALNS uses an adaptive layer in the step where the operators are selected, by using a score ϕ_j which measures the best performance of an operator j during the search to choose the most successful one. The adaptiveness lies in a roulette wheel mechanism which selects an operator j with a probability $\frac{\phi_j}{\sum_i \phi_i}$. During M iterations, the score ϕ_j is updated and the probabilities of selecting an operator are recalculated.

In order to design an ALNS algorithm for a given optimization problem we need to:

- Choose a number of fast construction operators which are able to construct a full solution.
- Select a number of destruction operators. It might be sufficiently important to choose the destruction operators that are expected to work well with the construction operator, but it is unnecessary.

Here is the detailed algorithm:

Algorithm 1 adaptive large neighborhood search

Construct a feasible solution x ; set $x^b = x$
repeat
  Choose a destroy neighborhood d and a repair neighborhood r using roulette wheel selection based on previously obtained scores π_j
  Generate a new solution x^t from x using the heuristics corresponding to the chosen destroy and repair neighborhoods
  if x^t can be accepted **then**
    $x = x^t$
  end if
  if $c(x^t) < c(x)$ **then**
    $x^b = x^t$
  end if
  Update scores π_j of d and r
until Stop criteria is met **return** x^b

Initial solution generation In order to deal with the initial solution, we apply a greedy algorithm, which will be used in the reconstruction phase of the ALNS. This operator aims to insert the non-inserted nodes by testing the different possible configurations and then giving a feasible solution. It is not necessary that the completion time of the initial solution be minimal, as this solution will be further enhanced using the ALNS method.

Solution destruction During the destruction phase, we put forward three different removal methods to maintain the diversity during the searching process and to define the neighborhood to explore at each iteration. Each removal method aims to remove a predefined number of nodes.

The first operator is known as proximity operator. Its objective is to select close clients according to a spatio-temporal measure [41], and then remove clients engendering the higher value of this measure.

Using the same technique, the route portion operator comes to give more flexibility than the proximity operator to change the routes. The principle consists in choosing a pivot client owned to a road and remove it as well as its adjacent clients. Then, we calculate the spatio-temporal measure, with the objective to select a second client belonging to another route but close to the initial pivot. This second pivot is removed from the solution as well as its adjacent clients and so on until all clients will be removed.

The third operator is referred to as longest detour operator. The interest of this operator is to remove the customers that lead to the largest cost increases for servicing them. For more details, we refer the reader to [42].

Solution reconstruction Solomon's insertion heuristic (1987) presented a technique for choosing the new customer to be inserted into a route using two criteria $c_1(i, u, j)$ and $c_2(i, u, j)$ to select customer u for insertion between adjacent clients i and j in the current partial route. The primary criterion c_1 calculates the best feasible insertion place in the current route for each unrouted client u as

$$c_1(i(u), u, j(u)) = \min_{\rho=1,\dots,m} c_1(i_{\rho-1}, u, i_\rho) \tag{17}$$

The second criterion c_2 selects the new inserted customer:

$$c_2(i(u^*), u^*, j(u^*)) = \max_u \{c_2(i(u), u, j(u)) | u \text{ is unrouted and route is feasible}\}$$

Customer u^* is then inserted into the route between $i(u^*)$ and $j(u^*)$. The measurement of an insertion place c_1 depends on two factors: the increase in total distance of the current route after the insertion, and the delay of service start time of the customer following the new inserted customer. To be more precise, $c_1(i, u, j)$ is calculated as:

$$c_1(i, u, j) = \alpha_1 \ (d_{iu} + d_{uj} - \mu \ d_{ij}) + \alpha_2 \ (b_{ju} - b_j), \tag{18}$$

$$\alpha_1 + \alpha_2 = 1, \quad \alpha_1 \geq 0, \quad \alpha_2 \geq 0, \quad \mu \geq 0, \tag{19}$$

where $d_{iu} + d_{uj}$ is the new distance between two nodes i and j after the insertion, b_j is the previous service start time, d_{ij} is the old distance between i and j and b_{ju} is the new service start time of customer j after the insertion of customer u. The criterion $c_2(i, u, j)$ is calculated as following

$$c_2(i, u, j) = \lambda \ d_{0u} - c_1(i, u, j), \lambda \geq 0 \tag{20}$$

where the parameter λ is used to define how much the best insertion place for an unrouted customer depends on its distance from the depot and extra time required to visit the customer by the current vehicle.

Roulette wheel selection For each iteration of the destruction phase, a roulette-wheel procedure is applied to select a method for generating the neighborhood (nodes to be removed). During the search process, the ALNS maintains a score φ_j which measures the best performance of an heuristic j in the past iterations. The roulette wheel selection consists in selecting an heuristic j with a probability $\frac{\varphi_j}{\sum_i \varphi_i}$. During M iterations, the score φ_i is reset and the probabilities of choosing an heuristic are recalculated [39].

4 Numerical Experiment

To investigate the performance of the ALNS in the context of the considered VRPFLexTW, we accomplished several computational tests. The algorithm was examined on a group of small instances in reference to the benchmark of Solomon, 1987, and its extension the instances of Gehring and Homberger, 1999. The solomon set R containing randomized customers is used. We applied the algorithm on a number of Solomon's Sizes as a benchmark example. he algorithms were implemented in Java 7, compiled with Intel compiler Celeron 1.80 GHz core i5 with 8GB RAM. The MALNS approach was run for 15600 iterations and was applied 10 times to each instance.

The results are showcased in the tables below. Table 1 showcases the optimal values of the Vehicule routing cost obtained through an ALNS optimisation. Table 2 shows the optimal number of vehicles in the solution obtained for each client configuration. Lastly, Table 3 enumerates the execution time required to reach optimal convergence in each instance.

From the performance tables, we can see that the ALNS approach reaches lower cost values in comparison with the Ant Colony Optimiser. Moreover, the quality of solutions obtained employs less vehicles to serve all target clients and take less time

Table 1 Value of the cost function for different numbers of clients

Solomon size	ALNS	ACO
100-client	1640	2635
200-client	4846	11074
400-client	12370	31702
600-client	26785	71154
800-client	51281	133482
1000-client	85904	219890

Table 2 Optimal number of vehicles corresponding to each configuration of clients

Solomon size	ALNS	ACO
100-client	10	27
200-client	12	65
400-client	24	141
600-client	40	224
800-client	65	307
1000-client	93	398

Table 3 Execution time of the ALNS Algorithm in seconds corresponding to each clients configuration

Solomon size	ALNS	ACO
100-client	0.96	2.47
200-client	3.86	7.9
400-client	34.41	34.2
600-client	168.01	85.53
800-client	489.19	187.36
1000-client	1086.41	263.75

to compute. This superior performance is due to the nature of the ALNS search which is adaptive and scales well with complexity. In contrast, Ant Colony is population based therefore the resource consumption is really heavy on memory and computing units.

5 Conclusion

Our main goal in this chapter was to review the ALNS approach for the VRPFLexTW problem. The state of the art concerning the VRPFlexTW is laid out and the Adaptive Large Neighbourhood Search is described as well. A comparison between the Ant colony optimizer and the ALNS approach shows the superiority of the latter for these types of problems due to its interesting mechanism of construction and deconstruction operator that are capable of attaining quality solutions in shorter execution times and with less computational overhead.

References

1. C. Dhaenens, M.L. Espinouse, B. Penz, Problèmes combinatoires classiques In Recherche opérationnelle et réseaux : méthodes d'analyse spatiale. Hermès Science Publications (2002)
2. G. Dantzig, R. Fulkerson, S. Johnson, Solution of a large-scale travelling salesman problem. J. Oper. Res. Soc. Am. **2**(4), 393–410 (1954)
3. R. Baldacci, A. Mingozzi, R. Roberti, Recent exact algorithms for solving the vehicle routing problem under capacity and time window constraints. Eur. J. Oper. Res. **218**(1), 1–6 (2012)
4. S.R. Balseiro, I. Loiseau, J. Ramonet, An ant colony algorithm hybridized with insertion heuristics for the timedependent vehicle routing problem with time windows. Comput. Oper. Res. **38**(6), 954–966 (2011)
5. R. Baños, J. Ortega, C. Gil, A. Fernández, F. de Toro, A simulated annealing-based parallel multi-objective approach to vehicle routing problems with time windows. Expert Syst. Appl. **40**(5), 1696–1707 (2013)
6. K. Braekers, K. Ramaekers, I.V. Nieuwenhuyse, The vehicle routing problem: state of the art classification and review. Comput. Indust. Eng. **99**, 300–313 (2016)
7. N.A. El-Sherbeny, Vehicle routing with time windows: an overview of exact, heuristic and metaheuristic methods. J. King Saud Univ. (Sci.) **22**(3), 123–131 (2010)
8. E. Teymourian, V. Kayvanfar, G.H.M. Komaki, M. Zandieh, Enhanced intelligent water drops and cuckoo search algorithms for solving the capacitated vehicle routing problem. Inf. Sci. **334**(335), 354–378 (2016)
9. T. Vidal, Technical note: split algorithm in o(n) for the capacitated vehicle routing problem. Comput. Oper. Res. **69**, 40–47 (2016)
10. C. Koc, T. Bektaş, O. Jabali, G. Laporte, Thirty years of heterogeneous vehicle routing. Eur. J. Oper. Res. **249**(1), 1–21 (2016)
11. Geoff Clarke, John W. Wright, Scheduling of vehicles from a central depot to a number of delivery points. Oper. Res. **12**(4), 568–581 (1964)
12. G. Gutin, A.P. Punnen (eds.), *The Traveling Salesman Problem and Its Variations*, vol. 12. (Springer Science & Business Media, 2006)
13. P. Toth, D. Vigo, An overview of vehicle routing problems. Veh. Routing Probl. 1–26 (2002)
14. J.K. Lenstra, A.H.G.R. Kan, Complexity of vehicle routing and scheduling problems. Networks **11**(2), 221–227 (1981)
15. N. Azi, M. Gendreau, J.Y. Potvin, An exact algorithm for a vehicle routing problem with time windows and multiple use of vehicles. Eur. J. Oper. Res. **202**(3), 756–763 (2010)
16. B. Eksioglu, A.V. Vural, A. Reisman, The vehicle routing problem: a taxonomic review. Comput. Indust. Eng. **57**, 1472–1483 (2009)
17. K. Braekers, K. Ramaekers, I. Nieuwenhuyse, The vehicle routing problem: state of the art classification and review. Comput. Ind. Eng. **99** (2015)
18. T. Vidal, G. Laporte, P. Matl, A concise guide to existing and emerging vehicle routing problem variants. Eur. J. Oper. Res. **286**, 401–416 (2019)
19. O. Bräysy, M. Gendreau, Vehicle routing problem with time windows, part ii: metaheuristics. J. Transp. Sci. **39**, 119–139 (2005)
20. B. Kallehauge, J. Larsen, O.B. Madsen, M. Solomon (eds.), Vehicle routing problem with time windows. *Column Generation*, pp. 1:67–98 (2005)
21. H. Ben Ticha, N. Absi, D. Feillet, A. Quilliot, Empirical analysis for the vrptw with a multigraph representation for the road network. Comput. Oper. Res. **88**:103–116 (2017)
22. O. Bräysy, M. Gendreau, Tabu search heuristics for the vehicle routing problem with time windows. Top J. **10**, 211–237 (2002)
23. J.Y. Potvin, T. Kervahut, B.L. Garcia, J.M. Rousseau, The vehicle routing problem with time windows part i: tabu search. INFORMS J. Comput. **8**, 158–164 (1996)
24. É. Taillard, P. Badeau, M. Gendreau, F. Guertin, J.Y. Potvin, A tabu search heuristic for the vehicle routing problem with soft time windows. Transp. Sci. **31**, 170–186 (1997)

25. S. Afifi, D. C. Dang, A. Moukrim, A simulated annealing algorithm for the vehicle routing problem with time windows and synchronization constraints. in *International Conference on Learning and Intelligent Optimization*, Lecture Notes in Computer Science, vol. 7997 (2013)
26. W.C. Chiang, R.A. Russell, Simulated annealing metaheuristics for the vehicle routing problem with time windows. Ann. Oper. Res. **63**, 3–27 (1996)
27. S. Røpke, D. Pisinger, An adaptive large neighborhood search heuristic for the pickup and delivery problem with time windows. Transp. Sci. **40**, 455–472 (2006)
28. H. Gehring, J. Homberger, A parallel hybrid evolutionary metaheuristic for the vehicle routing problem with time windows. in *Proceedings of EUROGEN'99*, pp. 57–64 (1999)
29. S. Thangiah, Vehicle routing with time windows using genetic algorithms. in *Application Handbook of Genetic Algorithms: New Frontiers*, pp. 253–277 (1995)
30. L. M. Gambardella, E. Taillard, G. Agazzi, Macs-vrptw: A multiple ant colony system for vehicle routing problems with time windows. in *New Ideas in Optimization*, pp. 63–76 (1999)
31. X. Tan, X. Zhuo, J. Zhang, Ant colony system for optimizing vehicle routing problem with time windows (vrptw). in *Computational Intelligence and Bioinformatics. ICIC 2006*. Lecture Notes in Computer Science, vol. 4115, pp. 33–38 (2006)
32. A. Dhahri, A. Mjirda, K. Zidi, K. Ghedira, A vns-based heuristic for solving the vehicle routing problem with time windows and vehicle preventive maintenance constraints. Procedia Comput. Sci. **80**, 1212–1222 (2016)
33. N. Mladenović, P. Hansen, Variable neighborhood search. Comput. Oper. Res. **24**, 1097–1100 (1997)
34. E.-G. Talbi, Combining metaheuristics with mathematical programming, constraint programming and machine learning. Ann. Oper. Res. **31**:170–186 (2013)
35. M. Solomon, Algorithms for the vehicle routing and scheduling problem with time window constraints. Oper. Res. J. **35**, 254–265 (1987)
36. S. Røpke, Palns-a software framework for parallel large neighborhood search. in *8th Metaheuristic International Conference CDROM, Metaheuristic International Conference* (Hamburg, Germany, 2009)
37. V.C. Hemmelmayer, Sequential and parallel large neighborhood search algorithms for the periodic location routing problem. Eur. J. Oper. Res. **243**, 352–60 (2014)
38. V. Pillac, M. Gendreau, C. Guéret, A.L. Medaglia, A parallel matheuristic for the technician routing and scheduling problem. Optim. Lett. **7**, 1525–1535 (2013)
39. D. Pisinger, S. Ropke, A general heuristic for vehicle routing problems. Comput. Oper. Res. **34**, 2403–2435 (2007)
40. S. Ropke, D. Pisinger, An adaptive large neighborhood search heuristic for the pickup and delivery problem with time windows. Transp. Sci. **40**(4), 455–472 (2006)
41. P. Shaw, Using constraint programming and local search methods to solve vehicle routing problems, in *International Conference on Principles and Practice of Constraint Programming* (Springer, Berlin, 1998), pp. 417–431
42. E. Prescott-Gagnon, G. Desaulniers, L.M. Rousseau, A branch-and-price-based large neighborhood search algorithm for the vehicle routing problem with time windows. Netw. Int. J. **54**(4), 190–204 (2009)

VRPTW: From Mathematical Models to Computing Science Tools

Nasri Mehdi, Metrane Abdelmoutaib, and Hafidi Imad

Abstract This paper provides effective methodologies intended to improve the behaviour of the Adaptive Large Neighborhood Search algorithm (ALNS), when solving the vehicle routing problem with time windows. The main challenge is to obtain a reduced real-time solution without compromising the quality of the solution specially for large instances. In this context, we present four distinct methodologies to deal with the problem on its deterministic and robust structure. In order to evaluate these methods, numerical experiments and comparisons using different instance sizes are provided.

Keywords VRPTW · ALNS · Parallel multi-threading · Robust approach

1 Introduction

Since the publication of the fundamental work of Dantzig, the Vehicle Routing Problem (VRP) has detonated drastically. The VRP has broadened its scope by introducing additional features with infer an enormous number of variations different from the one presented in [1] and summed up in [2]. For given clients with known positions and demands, the objective of the VRP problem in its overall structure is to plan a cluster of least expense routes that serve various geographically disseminated clients around the depot and satisfying explicit requirements of the problem. Actually, the problem becomes more complex when including some additional restrictions, such as the consideration of time windows constraints. We call this variant in the literature, Vehicle Routing Problem with Time Windows (VRPTW).

The VRPTW has been the subject of several pieces of research (e.g., [3, 4]), the goal is to serve all the customers in a specific interval of time. Typically, there are two kinds of the problem such as soft time windows and hard time windows. In the hard

N. Mehdi (✉) · M. Abdelmoutaib · H. Imad
National School of Applied Sciences Khouribga, Sultan Moulay Slimane University, 23052 Beni-Mellal, Morocco
e-mail: mehdi.nasri@usms.ac.ma

A. Nachaoui et al. (eds.), *Mathematical Control and Numerical Applications*, Springer Proceedings in Mathematics & Statistics 372,
https://doi.org/10.1007/978-3-030-83442-5_2

version of the problem, the time windows constraints must be respected, if the vehicle arrives at a customer earlier than expected, it must wait until the customer opens and it is also not allowed to arrive late. Nevertheless, the time windows restrictions may be violated taking into consideration additional penalty costs to the solution cost in the soft time windows.

A fascinating subject on tackling VRP constructed from considering parameters influenced by uncertainty, making the problem more sensible. Various methodologies have been proposed to handle uncertainties in a VRP problem either in demand, travel time, as well as service time (see for e.g. [5, 6]). Into the works, we recognize stochastic and robust methodologies. The stochastic method can be viewed as a procedure that aims at finding a near-best solution which responds to all uncertain parameters which are characterized by their probability distributions [7]. In the contrast, the reason for the robust methodology is to discover the solution which immunizes against the effect of data uncertainties, taking into account some measure challenges such as the worst case, best case, min-max deviation. The literature coats a large number of publications in this field, we can cite for instance [8, 9].

This paper considers the optimization of real world field vehicle routing problem with time windows. The aim is to find a set of least cost routes, minimizing a travel distance of vehicles, starting and ending from a terminal, to serve a few clients. We developed our research by investigating two categories of this problem: deterministic and robust problems. We describe a modified adaptive large neighborhood search (MALNS) conceived to tackle the vehicle routing problem with time windows (VRPTW). This technique integrates the modified choice function as selection mechanism in the ALNS method which appears well-suited for the VRPTW. Moreover, we illustrate a new multi-threading parallel adaptive large neighborhood search to tackle the same problem. More precisely, we applied a parallel computing to the greedy insertion involved in the initial block and the destroy/repair operators incorporated into the ALNS process. Our algorithm is shown to be able to obtain a good solution in a reasonable time without too much compromise on the quality of the solution.

The robust vehicle routing problem with time windows under both travel times and service times uncertainties is also considered. For this purpose, an effective resolution robust approach has been presented to minimize the total distance of the travel time in the presence of the maximum deviations of possible uncertain data. In this regard, we generated all possible scenarios by using Metropolis-Monte Carlo simulation and we opt for the adaptive large neighborhood search ALNS algorithm to solve each sub-problem related to each scenario. In order to study the feasibility of the resulting solution, efficient mechanisms have been developed; the first concerns the verification of the robustness, while the second takes into consideration the evaluation of the solution on the worst case.

As far as the adopted approach derives the best robust solution that responds to all uncertainties, it actually suffer from lengthy computational times; This is due to the fact that the generation of scenarios, as well as the ALNS block are time-consuming. As an option to remedy this problem, we present a procedure for the thread parallelism involved in the Monte Carlo block, and we use the parallel ALNS proposed by [10]. This gives rise to four different robust approaches combining the (sequential/parallel) Monte Carlo algorithm and the (sequential/parallel) ALNS. The considered approaches are tested on Solomon's benchmark instance of VRPTW and lager instances generated randomly. Accordingly, we can offer a decision-making solution that provides great protection against delays in a reasonable running time.

2 Mathematical Formulation

We consider for the Vehicle Routing Problem (VRP) the following objective function:

$$\mathbf{Min} \sum_{k \in V} \sum_{(i,j) \in A} d_{ij}\ x_{ij}^{k}$$

where the objective function $\sum_{k \in V} \sum_{(i,j) \in A} d_{ij}\ x_{ij}^{k}$ refers to the total traveled distance d_{ij} and x_{ij}^{k} are binary decision variables that take the value 1 if vehicle k which belongs to the set of vehicles V uses the edge (i, j) and zero otherwise. The set of arcs of different routes is denoted as A and contains pairs of nodes (i, j). The constrained optimization problem must be accomplished under some constraints:

$$\sum_{k \in V} \sum_{j \in N} x_{ij}^{k} = 1 \qquad \forall\, i \in N, \tag{1}$$

This first constraint stipulates that each customer must be visited once.

$$\sum_{j \in N} x_{0j}^{k} = 1 \qquad \forall\, k \in V, \tag{2}$$

The constraint (2) guarantees that each tour starts from the depot.

$$\sum_{i \in N} x_{ih}^{k} - \sum_{j \in N} x_{hj}^{k} = 0 \qquad \forall\, h \in N \quad \forall\, k \in V, \tag{3}$$

The constraint (3) is a flow conservation, and consequently ensures that the same vehicle arrives and leaves from each node it serves.

$$\sum_{i \in N} x_{i0}^{k} = 1 \qquad \forall\, k \in V, \tag{4}$$

This fourth constraint guarantees that each tour returns to the depot.

In order to forbid the violation of the time windows at any client, we require two additional constraints:

$$a_i \le P_i^k \le b_i \qquad \forall\, i \in N \quad \forall\, k \in V, \tag{5}$$

$$P_i^k + d_{ij} - P_j^k \le M(1 - x_{ij}^k) \qquad \forall\, i \in N \quad \forall\, j \in N \setminus \{0\} \quad \forall\, k \in V. \tag{6}$$

with M a great value.

It is worth pointing that the travel time t_{ij} is subject to uncertainty Δ_{ij}. The service time P_i^k depends on uncertainty δ_i. We identify the uncertainty sets corresponding to these two parameters by:

$$U_t = \{\tilde{t} \in \mathbb{R}^{|A|} \;/\; \tilde{t_{ij}} = t_{ij} + \Delta_{ij}\epsilon_{ij},\ \sum_{(i,j)\in A} \epsilon_{ij} \le \Gamma, 0 \le \epsilon_{ij} \le 1,\ \forall (i,j) \in A\}$$

and

$$U_P = \{\tilde{P} \in \mathbb{R}^{|N|} \;/\; \tilde{P_i} = P_i + \delta_i\omega_i, \sum_{i\in N} \omega_i \le \Lambda, 0 \le \omega_i \le 1,\ \forall i \in N\}$$

The objective function for the robust problem becomes:

$$\textbf{Min}\, (\sum_{k\in V}\sum_{(i,j)\in A} x_{ij}^k t_{ij} + \max_{\{\Psi/\Psi\subset A, |\Psi|=\Gamma\}} \sum_{k\in V}\sum_{(i,j)\in\Psi} x_{ij}^k \Delta_{ij})$$

And the Eq. (6) becomes:

$$P_i^k + t_{ij} + \delta_i \nu_i^\theta + \Delta_{ij}\mu_{ij}^\Psi - P_j^k \le (1 - x_{ij}^k)M,$$

$$\forall (i \in N),\ \forall (j \in N \setminus \{0\}),\ \forall (k \in V)\ \forall (\theta \subset N) \mid \theta \mid = \Lambda,\ \forall (\Psi \subset A) \mid \Psi \mid = \Gamma$$

where M is a great value and ν_i^θ and μ_{ij}^Ψ are two indicator functions. When $i \in \theta$, ν_i^θ takes the value of one. When $(i, j) \in \Psi$, μ_{ij}^Ψ takes one.

3 Adaptive Large Neighborhood Search

The ALNS is a metaheuristic proposed by Ropke and Pisinger [11] as a development of the Large Neighborhood Search (LNS) heuristic presented by Shaw [12]. It is an adaptive approach employed to ameliorate an incumbent solution. The basic idea of this agorithm is to improve a given initial solution by exploring many large neighborhoods. This can be done by the application of various destroy and repair operators. More precisely, the destroy operator removes nodes from the solution. We obtain then an incomplete or infeasible solution d(x). The repair operator reinserts the removed nodes at more favored position which leads to a new feasible (complete) solution r(d(x)). The resulting solution will be accepted or rejected based on a Hill Climbing acceptance criteria which only accepts solutions that are better than the current one. This process will be terminated when a stop criterion is met.

It is worth mentioning that ALNS uses an adaptive layer in the step where the operators are selected, by using a score ϕ_j which measures the best performance of an operator j during the search to choose the most successful one. The adaptiveness lies in a roulette wheel mechansim which selects an operator j with a probability $\frac{\phi_j}{\sum_i \phi_i}$. During M iterations, the score ϕ_j is updated and the probabilities of selecting an operator are recalculated.

Here is the detailed ALNS algorithm:

Algorithm 1 Adaptive Large Neighborhood Search

Input: Initial solution x
Parameters: repair neighborhood N^+, destroy neighborhood N^-
Initialization phase: $x^b = x$, $\rho^+ = 1$ and $\rho^- = 1$
repeat
  Choose a destroy neighborhood N^- using roulette wheel selection based on previously obtained scores ϕ_j
  Apply the destroy heuristic corresponding to the destroy neighborhood $d(x)$
  Choose a repair neighborhood N^+ using roulette wheel selection based on previously obtained scores ϕ_j
  Apply the repair heuristic corresponding to the repair neighborhood $r(d(x))$
  if $r(d(x))$ can be accepted **then**
    $x = r(d(x))$
  end if
  if $c(r(d(x))) < c(x)$ **then**
    $x^b = r(d(x))$
  end if
  Update scores ϕ_j of d and r
until Stop criteria is met
return x^b

4 Deterministic Vehicle Routing Problem with Time Windows (VRPTW)

In this section, we put forward two different alternatives to enhance the ALNS process. The first approach is the modified adaptive large neighborhood search (MALNS) [13]. It consists on a framework of metaheuristic designed to solve the vehicle routing problem with time windows. The approach integrates a new mechanism to deal with the Adaptive Large Neighborhood Search (ALNS) algorithm applied to the problem. The major challenge is to highlight intensification over diversification within the heuristic search process. In this context, we incorporate the process of the choice function proposed by [14] into the ALNS algorithm. In this way, a few destroy/repair methods are combined to explore multiple neighborhoods within the same search and defined implicitly the large neighborhood. In the second alternative, we provide a new approach of parallel adaptive large neighborhood search algorithm designed to solve the vehicle routing problem with time windows. The main challenge is to obtain a reduced real-time solution without compromising the quality of the solution specially for large instances. Our main contribution introduces a procedure for the Thread parallelism of greedy insertion used in the initial block and also the destroy/repair operators involved in process of the ALNS metaheuristic. For this cheapest insertion Solomon's greedy heuristic, we refer the reader to the paper of ([15]).

4.1 *The Modified Adaptive Large Neighborhood Search (MALNS)*

The intention of the MALNS is to effectively investigate the search space using some large neighborhoods. The process consists on integrating the modified choice function of [14] into the selection level of the ALNS while sparing the whole process. In other terms, the destroy and repair operators will not be chosen by the roulette wheel mechanism as in the original ALNS, but they will be rather chosen by the MCF. Here is the detailed algorithm of the MALNS [13]:

Algorithm 2 Modified Adaptive Large Neighborhood Search

```
Construct a feasible solution x ; set x^b = x
repeat
  Choose a destroy neighborhood d and a repair neighborhood r using the modified choice
  function based on the obtained scores F_t
  Generate a new solution x^t from x using the heuristics corresponding to the chosen destroy and
  repair neighborhoods
  if x^t can be accepted then
    x = x^t
  end if
  if c(x^t) < c(x) then
    x^b = x^t
  end if
  Update scores F_t of d and r
until Stop criteria is met
return x^b
```

For reasons of completeness, we will explain the initialization step as well as the removal and insertion operators involved in the destroy/repair block to guarantee the diversity during the searching process.

In order to handle the initial solution, we used the greedy insertion heuristic introduced by [16]. This method consists of finding the best location of a given node by testing the different possible configurations. More specifically, the algorithm selects the best possible insertion place in the present route for each non inserted node tacking into account two parameters: the increase in total cost of the present route after the insertion, and the delay of service start time of the client following the new inserted client. This process ends when all deleted nodes will be inserted.

During the destruction phase, we adopt three different destruction operators. The first operator is called proximity operator, it aims to remove a set of customers which are similar in accordance with the spatio-temporal measure ([12]). Similarly, the route portion operator gives more flexibility of change on the routes by choosing a pivot client attributed to a route and delete it with its adjacent. The third operator is the longest detour operator. It tries to delete the customers that lead to the largest increase of the cost of the current solution.

After the destruction phase, the obtained solution must be repaired so as to get a complete solution. Therefore, we use two repair operators. First, the greedy insertion [16] tries to select the location that reduces the insertion cost over all nodes and routes. Second, the regret insertion ([17]) defines a regret value which is the cost difference of inserting the customer i in its best route and its second best route. Thereby, customers with the highest value should be inserted first.

4.2 The Multi-threading Parallel ALNS

In this section, we propose a simplified presentation of our algorithm with three levels developed in one or more algorithms illustrated in Fig. 1. At the first level or the initialization phase, we use a parallel greedy insertion to produce an initial feasible solution. The principle of this greedy insertion is to select the best feasible insertion place in the current route for each non inserted node considering two factors: the increase in total cost of the current route after the insertion, and the delay of service start time of the client following the new inserted client.

After this initialization phase, we split the process in three steps. The first step is devoted to removing a fixed number of nodes such that each thread contributes to rip up a part of the current initial solution obtained during the greedy insertion processing by using three layers of destruction operators. After a complete run of this metaheuristic algorithm, we aggregate in the next processing step the formed destroy solution and send it back to the master processor. In the meanwhile, each thread worker reads different route file and solve set covering problem instances by repairing the solution based on the Solomon's insertion heuristic which we will explain later in more detail in the next section. In the last step, we combine routes from different solutions obtained by distinct threads into repaired solution. At the end, a new solution is accepted or rejected according to an acceptance criteria of Hill Climbing type. This process will be repeated until a stop criterion is met. We point out that Hill climbing is at the extreme of the spectrum of acceptance criteria, which only accepts solutions that improve on the current one.

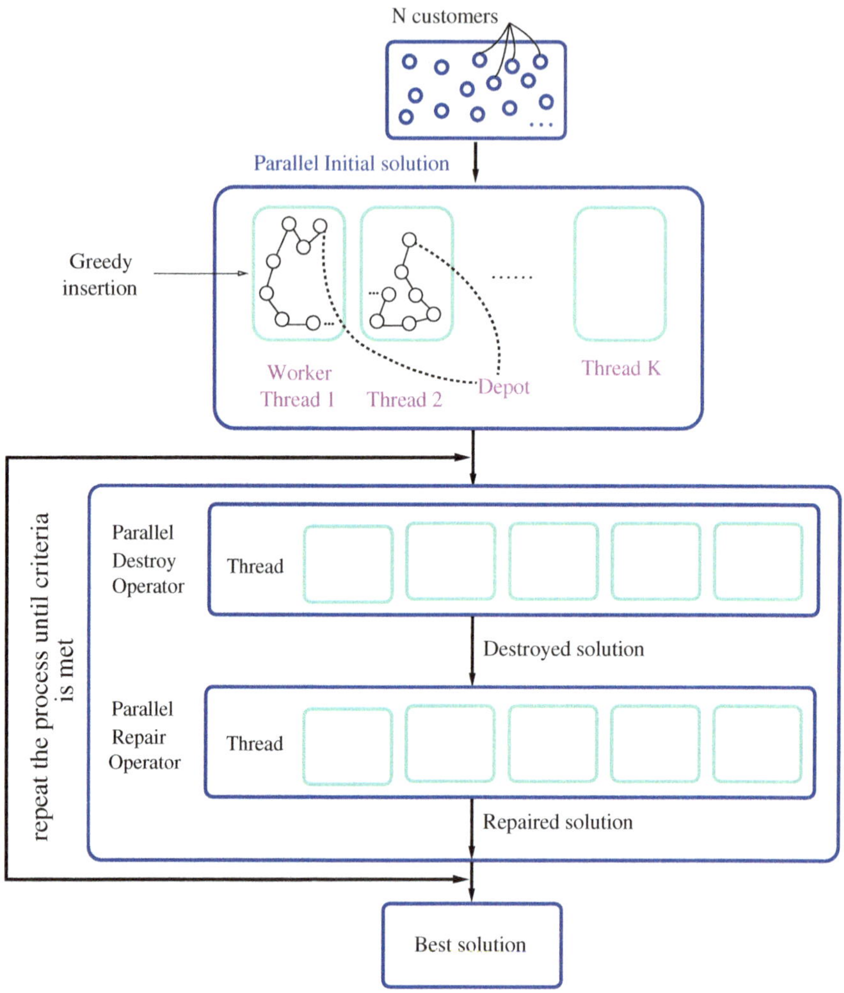

Fig. 1 The parallel adaptive large neighborhood search

5 Robust Vehicle Routing Problem with Time Windows (RVRPTW)

In this section, we study a variant of the Robust Vehicle Routing Problem with Time Windows (RVRPTW) including both uncertainties in travel and service times. The advantages of the used methods lie in the introduction of an effective way of modeling uncertain data, the choice of a mathematical model and the methods to evaluate that robust solution that can solve the whole problem, and also the selection of an adaptive large neighborhood search heuristic to solve each problem related to each scenario [8].

5.1 Sequential Approach to Solve the RVRPTW

An iteration of the Robust approach algorithm starts with the generation of a set of deterministic sub-problems called realizations (by Monte-Carlo Simulation). Each realization $\omega_N^{\Lambda,\Gamma}$ represents a possible scenario in which the displacement times associated with a subset of arcs $\Psi \subset A$ of cardinality Γ attain their maximum values $t_{ij} = t_{ij} + \Delta_{ij}$, and service times of a subset of vehicles $\theta \subset V$ of cardinality Λ attain their maximum values $P_i = P_i + \delta_i$. While the other arcs and the other nodes conserve their nominal values t_{ij} and P_i.

For each realization $\omega_N^{\Lambda,\Gamma}$, we apply the Adaptive large neighborhood search in order to get a feasible solution mentioned Sol_N which satisfies each scenario that we have already generalized.

Even if the solution Sol_N is realizable on the scenario $\omega_N^{\Lambda,\Gamma}$, it may violate window time constraints if it considers other realizations. Therefore, we need to check the robustness of the solution without the need to test it on all possible realizations. Definitely, a solution Sol_N is robustly feasible if all its tours obey the time windows constraints at any node visited, where at most, Γ displacement times and Λ service times are subject of uncertainty.

In the last step, we evaluate the solution Sol_N on the worst of possible cases, which corresponds to the realization where the displacement times associated to the Γ arcs with the worst deviations and belonging to this solution, reach simultaneously their maximum values. First, we put in descending order all the arcs of the solution according to their maximum deviations. Then, we assign to the first Γ arcs their maximum displacement time and the remaining arcs nominal displacement time. Finally, we carry out a summation of the times obtained to determine the worst-case evaluation of cases. For more details, we refer the reader to the paper of [8].

5.2 Multi-threading Parallel Approach to Solve the RVRPTW

In this section, we study the effect of multi-threading parallelization of the sequential resolution approach blocks on the running time and the objective function. The contribution to the previous work lies first on the choice of the efficient Parallel Adaptive Large Neighborhood Search (PALNS) metaheuristic of [10], which leads to a reduced running time. Moreover, we used our parallel version of the Metropolis Monte Carlo algorithm to generate all possible realizations and to transform the problem under uncertainties to a set of deterministic sub-problems. For more detail about this splitting process, see for instance the previous work [8] and the references therein. Based on the efficient implementation of [10], different combinations (sequential/parallel) of the Monte Carlo algorithm and ALNS are performed. In this way, our strategy offers to decision makers the choice of the combination depending on their preferences and the situation at hand. For more details, we refer the reader to the paper of [18].

6 Computational Experiments

In this section, we summarize some of the results obtained by evaluating different approaches designed to solve both the deterministic vehicle routing problem with time windows and the vehicle routing problem with time windows with uncertain service and travel times. All the algorithms were examined on a group of instances in reference to the benchmark of [16], and its extension the instances of [19], Notwithstanding, instances larger than 1000 are generated randomly so that the euclidian distance between each pairs of nodes goes from 0 to 100, the same thing for service time. The time interval is of capacity 200 between the start and the end of the operations at each node:

- Set R contains problems with randomized customers.
- Set C contains problems with clustered customers.
- Set RC contains problems with both randomized and clustered customers.

With a view to simulate the uncertainty parameters by discrete scenarios, we generate the uncertain travel time and uncertain service time randomly, so that they go from 0 to 10. We mention the used instances as follows: Gr_Γ_Λ, where Γ and Λ are respectively the number of travel times and service times considered uncertain and Gr relates to the instance of Solomon and Gehring and Homberger's benchmark or the size of larger instances.

6.1 Computational Results of the MALNS

Table 1 represents the MALNS improvement results on the basis of the objective value for some instance groups compared to the ALNS algorithm. We can remark

Table 1 Comparison of objective functions between the ALNS and MALNS

Instance	Initial solution	ALNS solution	MALNS solution
C121	2807.20	2704.57	2704.57
C221	2063.76	1931.44	1931.44
RC141	9652.33	9127.15	8724.36
RC241	7949.65	7471.01	7003.61
R161	23027.04	22838.65	22145.03
R261	22210.38	21945.30	19806.15
C181	25774.19	25184.38	25184.38
C281	12212.17	11687.06	11687.06
RC1101	49389.09	48702.83	47530.42
RC2101	35908.40	35073.70	33104.52

Table 2 Comparison of runtime in seconds between the ALNS and MALNS

Instance	ALNS	MALNS	Gap (%)
C121	435	504	15.86
C221	505	581	15.04
RC141	501	564	12.57
RC241	718	801	11.55
R161	556	606	8.99
R261	749	816	8.94
C181	708	743	4.94
C281	937	983	4.90
RC1101	903	921	1.99
RC2101	1331	1357	1.95

that the MALNS method exceeds the ALNS method, and provides better results in view of solution quality, the gain average percent can reach 10.36%.

Table 2 illustrates the execution time of our method compared to ALNS. We computed the gain average percent (GAP) which can be defined by the absolute difference between execution time of the MALNS and the ALNS divided by the magnitude of the execution time of the ALNS. It is clear that our MALNS takes more time of execution compared to the ALNS, with a percentage gap of an average between 22.84% and 1.95%. This can be explained by the fact that the choice function takes more time to choose the appropriate operator to use, in the contrast of the roulette wheel selection.

6.2 *Computational Results of the Multi-threading Parallel ALNS*

Table 3 presents the execution time of our new approach for different instance size depending on the number of threads in use. The choice of four threads is not restrictive and we can use threads as much as possible. The conclusion from this table is clear: More threads lead to the improvement of the average execution time.

Table 4 presents the objective function of thread parallelism approach compared to sequential ALNS. We computed then the mean absolute deviation (MAD) which is the absolute difference between cost function of the thread parallelism and the sequential ALNS divided by the magnitude of the objective function in sequential ALNS.In fact, when we divide the clients on the threads in the insertion phase, we reduce the chances of falling on a good neighbor and we increase the objective function of our parallel approach.

Table 3 The average of execution time of an ALNS iteration in (ms)

Instance size	1 thread	2 threads	3 threads	4 threads
100	28	25	23	15
1000	84	67	62	50
2000	219	137	120	117
3000	393	271	234	206
4000	819	580	350	305
5000	1049	667	471	421
6000	1954	989	667	570
7500	2992	1412	924	781
8000	3314	1547	985	816

Table 4 Sequential and multi-threading parallel ALNS comparison according to the objective function

Instance size	Sequential ALNS	Parallel ALNS	Mean absolute deviation
100	2252	2883	0.28
1000	17502	19777	0.13
2000	34420	36829	0.07
3000	50817	53866	0.06
4000	67675	71728	0.06
5000	82937	86249	0.04
6000	99072	102040	0.03
7000	116911	119241	0.02
8000	132631	133950	0.01

6.3 *Computational Results of the Sequential Robust Approach*

The Table 5 shows the results obtained for large instances by comparing the best-known results for the deterministic VRPTW to the results found by the robust approach based on ALNS that deals with the VRPTW considering that travel times and service times are both uncertain. The robust approach provides great protection against delays with a marginal increase in travel times and service times.

Table 5 Performance of our robust approach versus best known results

Instance	Initial solution	Best	Mean	Best known
R121_25_25_200	6610.35	5696.41	5911.84	4784.11
C121_25_25_200	3399.45	2915.86	3029.34	2704.57
R141_25_25_400	12526.98	10939.12	11082.53	10372.31
C141_25_25_400	9205.81	8138.18	8584.48	7152.02
R161_50_50_600	26176.96	24277.04	24910.6	21131.09
C161_50_50_600	17624.24	15511.06	16672.38	14095.64
R181_50_50_800	41205.13	39011.14	39643.9	36852.06
C181_50_50_800	29445.55	28491.39	29075.07	25030.36
R1101_100_100_ 1000	63493.03	60506.88	61702.03	53473.26
C1101_100_100_1000	52251.98	51066.63	51982.04	42478.95

6.4 Computational Results of the Parallel Robust Approaches

Table 6 puts forward a comparison of the execution time for each instance group between different robust approaches when the maximal number of iterations is fixed at 15,600. As intended, th results show that the approaches containing the parallel ALNS and those containing parallel Monte Carlo lead to the improvement of the average execution time compared to other sequential approaches.

Table 7 shows the objective function of different robust approaches for the group of instances 1000–4500. We remark that the cost function of the approaches containing parallel and sequential ALNS solution converges when the size of the instance is increasing. We caclculate the MAPD, which is the absolute difference between the cost function of the approach containing the sequential ALNS and the parallel ALNS divided by the magnitude of the objective function in the approach with sequential

Table 6 The execution time of different robust approaches in seconds for 15,600 iterations

Instance	MC sequential and ALNS sequential	MC sequential and ALNS parallel	MC parallel and ALNS sequential	MC parallel and ALNS parallel
100_10_10	832.24	701.03	725.45	574.24
200_25_25	1007.41	853.11	883.25	669.33
400_25_25	1331.47	1037.98	1101.12	770.91
600_50_50	1402.50	1160.24	1230.64	858.38
800_50_50	1545.39	1302.85	1331.68	979.14
1000_100_100	2687.50	2062.25	2165.74	1340.49
1500_100_100	4519.79	3267.98	3665.37	2413.56
2000_100_100	5606.81	3953.18	4389.74	2736.11

Table 7 Comparison of different approaches according to the objective function

Instance	MC sequential and ALNS sequential	MC sequential and ALNS parallel	MC parallel and ALNS sequential	MC parallel and ALNS parallel
1000_100_100	18,612	21,031	18,642	21,075
1500_100_100	25,961	28,557	25,904	28,516
2000_100_100	36,420	38,969	36,408	38,943
2500_100_100	44,981	48,129	44,998	48,161
3000_100_100	52,817	55,986	52,836	56,014
3500_100_100	60,356	63,977	60,321	63,937
4000_100_100	67,675	71,059	67,642	70,986
4500_100_100	75,306	78,318	75,329	78,371

ALNS. This indicator varies between 13.05% f and 3.96% for the instance of size 4500. Our assertion from this results is that incorporating the parallel ALNS is very beneficial for large instances.

7 Conclusion

In this work, we explored the idea of developing efficient methods to solve two main problems that model many real world issues in the field of the vehicle routing problems. The solution proposed approaches include both sequential and parallel methods.

After presenting the studied problem and giving some insights of motivation behind our work in the first section, we exposed the Adaptive Large Neighborhood Search metaheuristic, which seems well-suited for the VRPTW, its power is manifested in the fact that each new solution is obtained by first removing a number of vertices, then re-inserting these vertices into the solution. ALNS was chosen because it outperforms other mono-objective algorithms applied to the same problem while keeping the simplicity and high performance that characterize local search algorithms. The paper included two sections where each, focused on one of the variants of the vehicle routing problem.

The third section presented new approaches conceived to enhance the process of the ALNS in terms of objective function and execution time. The first approach consists on incorporating the modified choice function (MCF) ([14]) into the selection phase of the ALNS while sparing the whole process. In other words, the destroy and repair operators will not be selected by the roulette wheel mechanism as in the original ALNS, but they will be rather selected by the MCF. The second approach took advantage of thread parallelism with a view to select a good solution in a reasonable time without too much compromising the quality of the solution. The principle is to introduce the Thread parallelism of greedy insertion used in the initial block and also the destroy/repair operators implicated in the mechanism of the ALNS.

The fourth section of this paper focused on a new variant of VRPTW incorporating service time as a source of uncertainty in addition to the travel times. The problem is called then, the Robust Vehicle Routing Problem with Time Windows (RVRPTW). We considered a sequential robust approach which requires efficient mechanisms to derive a best robust solution that responds to all uncertainties. After using this sequential approach which include generating all possible scenarios by using Monte Carlo simulation and using the adaptive large neighborhood search ALNS algorithm to solve each sub-problem related to each scenario, a new thread parallelism strategy was introduced in those blocks. This leads to four different robust approaches combining the (sequential/parallel) Monte Carlo algorithm and the (sequential/parallel) ALNS.

The perspective of this work is to add a clustering step as a preprocessing treatment, and study its impact of the by comparing different cluster algorithms following similarity pair of clusters by introducing the linkage metrics (Minkowski, Mahanalobis, Pearson correlation, ...) for measuring proximity between clusters. Our guess is that this study will lead to improve our thread parallelism approach.

References

1. G. Dantzig, J. Ramser, The truck dispatching problem. Manag. Sci. **6**, 80–91 (1959)
2. G. Clarke, J.R. Wright, Scheduling of vehicle routing problem from a central depot to a number of delivery points. Oper. Res. **12**, 568–581 (1964)
3. O. Bräysy, M. Gendreau, Vehicle routing Problem with time windows. Part II: metaheuristics. J. Transp. Sci. **39**, 119–139 (2005)
4. B. Kallehauge, J. Larsen, O.B. Madsen, M.M. Solomon, Vehicle routing problem with time windows. Column Gener. **1**, 67–98 (2005)
5. M. Dror, P. Trudeau, Stochastic vehicle routing with modified savings algorithm. Eur. J. Oper. Res. **23**, 228–235 (1986)
6. M. Dror, G. Laporte, F.V. Louveaux, Vehicle routing with stochastic demands and restricted failures. ZOR-Model. Oper. Res. **37**, 183–273 (1993)
7. M. Gendreau, G. Laporte, R. Séguin, Stochastic vehicle routing. Eur. J. Oper. Res. **88**, 3–12 (1996)
8. M. Nasri, A. Metrane, I. Hafidi, A. Jamali, A robust approach for solving a vehicle routing problem with time windows with uncertain service and travel times. Int. J. Ind. Eng. Comput. **11**, 1–16 (2020)
9. L. Wu, M. Hifi, H. Bederina, A new robust criterion for the vehicle routing problem with uncertain travel time. Comput. Ind. Eng. **112**, 607–615 (2017)
10. S. Røpke, PALNS-A software framework for parallel large neighborhood search. in *8th Metaheuristic International Conference CDROM, Metaheuristic International Conference* (Hamburg, Germany, 2009)
11. S. Røpke, D. Pisinger, An adaptive large neighborhood search heuristic for the pickup and delivery problem with time windows. Transp. Sci. **40**, 455–472 (2006)
12. P. Shaw, Using constraint programming and local search methods to solve vehicle routing problems. in *International conference on principles and practice of constraint programming* (Springer, Berlin, 1998), p. 1520

13. M. Nasri, A. Metrane, I. Hafidi, A modified ALNS algorithm for vehicle routing problems with time windows. J. Phys. Conf. Ser. **1743**, (2021)
14. J. Drake, E. Özcan, E. Burke, An improved choice function heuristic selection for cross domain heuristic search. Lect Notes Comput Sci **7492**, 307–316 (2012)
15. O. Bräysy, M. Gendreau, Vehicle routing problem with time windows. Part I: route construction and local search algorithms. J. Transp. Sci. **39**, 104–118 (2005)
16. M. Solomon, Algorithms for the vehicle routing and scheduling problem with time window constraints. Oper. Res. J. **35**, 254–265 (1987)
17. M. Diana, M. Dessouky, A new regret insertion heuristic for solving large-scale dial-a-ride problems with time windows. Transp Res Part B Methodol **38**, 539–557 (2004)
18. M. Nasri, I. Hafidi, A. Metrane, Multithreading parallel robust approach for the VRPTW with uncertain service and travel times. Symmetry **13**, 36 (2020)
19. H. Gehring, J. Homberger, A parallel hybrid evolutionary metaheuristic for the vehicle routing problem with time windows. in *Proceedings of EUROGEN'99*, pp. 57–64 (1999)

An Anisotropic PDE for Multi-frame Super-Resolution Image Reconstruction

Idriss El Mourabit, Abdelilah Hakim, and Amine Laghrib

Abstract The principal idea of the multi-frame super-resolution (SR) is to generate a higher resolution (HR) image from a set of lower resolution (LR) ones. Since, SR techniques suffers in particular from restoration artifacts such as staircasing effect, noise and blurring of edges. In this work, we tried to enhances the performance of super-resolution by using an anisotropic PDE based a structure tensor. To define this tensor, we takes the benefit from the classical model of Perona-Malik that allow readily to select the image homogeneous regions and use nonlinear tensor of Weickert near edges. Then, the proposed equation can preserve more the image features (corners and edges) much better while avoiding restoration artifacts. The efficiency and robustness of the our approach are illustrated and compared to other methods through numerical simulations.

Keywords Multi-frame super-resolution · SR · Image restoration · Partial differential equations · Tensor diffusion

1 Multi-frame SR Model

In recent years, the multi-frame super-resolution reconstruction is one of the most important inverse problems in image processing. The SR techniques play a significant role in a variety of scientific areas such as medical imaging, astronomical imaging and anatomy. However, HR images are not always obtainable. Indeed, the setup of high-resolution imaging is costly because of intrinsic limitations of the sensor, optics

I. El Mourabit (✉)
Ecole Supérieure de Technologie d'Essaouira, Ghazoua, Morocco

A. Hakim
Faculty of Science and Technology of Marrakech, Department of Mathematics and Informatics, University Cadi Ayyad, Marrakech, Morocco

A. Laghrib
LMA FST Béni-Mellal, Université Sultan Moulay Slimane, Beni-Mellal, Morocco

A. Nachaoui et al. (eds.), *Mathematical Control and Numerical Applications*,
Springer Proceedings in Mathematics & Statistics 372,
https://doi.org/10.1007/978-3-030-83442-5_3

manufacturing technology. These constraints can be overcome through the use of SR techniques, that are relatively inexpensive.

In practice, the acquired images of a real scene are typically of low resolution due to some degradation operators. In particular, the observed images are corrupted by noise, suffered from blurring and decimated. Basically, we suppose that the set of low-resolution (LR) images is acquired by the same sensor and taken in the same environmental conditions. The high-resolution (HR) image $\mathbf{u}$ (represented as a vector of size $\left[q^2n^2 \times 1\right]$, with q is the factor of resolution enhancement) is related to the low-resolution (LR) images $\mathbf{v_k}$ (represented as a vector of size $\left[n^2 \times 1\right]$), by the following formula

$$\mathbf{v_k} = DF_k H\mathbf{u} + \eta_k \quad \forall k = 1, 2, \ldots, n$$

where
H: represents the blurring operator of size $[q^2n^2 \times q^2n^2]$.
D: the matrix of decimation of size $[n^2 \times q^2n^2]$.
F_k: is a geometric warp matrix of size $[q^2n^2 \times q^2n^2]$, representing a non-parametric transformation that differs in all frames.
η_k: is a vector of size $[n^2 \times 1]$ represents the additive noise for each image.

The multi-frame super resolution process consists of reconstructing a version of the original image $\mathbf{u}$ from the set of low-resolution ones v_k, $k = 1, \ldots, n$. Due to the presence of the different degradation operators, the SR problem is difficult and ill-posed (non-uniqueness of solution). So,to reconstruct the image u we follow the approach of Farsiu et al. [1], it suggests to separate the SR into three steps:

- Computing the geometric warp matrix F_k for each image.
- Fusing the sequence of LR images $\mathbf{v_k}$ into a noised and blurred HR one $\mathbf{u_0} = H\mathbf{u}$
- Reconstructing the HR image $\mathbf{u}$ from the noised blurring one $\mathbf{u_0}$.

In the following we will focus only on the last step which is a restoration step.

We will compute the HR image $\mathbf{u}$ through a deblurring problem of the image $\mathbf{u_0}$, obtained from the fusion step. Unfortunately, this inverse problem is ill-posed in presence of noise and blur. To cope with this problem, we give some prior knowledge on the HR image $\mathbf{u}$ in a Bayesian framework. The image $\mathbf{u}$ has been known in the presence of Gaussian noise, the measured vector $\mathbf{v_k}$ is also a Gaussian one. Using the rule of Bayes, estimating the HR image $\widehat{\mathbf{u}}$ is equivalent to resolve a optimization problem by using the maximum a posteriori (MAP) super-resolution problem.

$$\begin{aligned}\widehat{u} &= \arg\max_u\{p(u/\mathbf{u_0})\}\\ &= \arg\max_u\{p(\mathbf{u_0}/u).p(u)\}\\ &= \arg\min_u\{-\log(p(\mathbf{u_0}/u)) - \log(p(u))\}\end{aligned} \tag{1}$$

with $p(\mathbf{u_0}/u)$ is the fidelity term (i.e. the relationship between the the HR image and the observed one), while $p(u)$ represents the prior knowledge imposed on the high-resolution image u. Using these two terms, we can present the general super-resolution problem as follows:

$$\widehat{u} = \underset{\mathbf{u}}{\operatorname{argmin}} \{\|Hu - \mathbf{u_0}\|_2 + \delta L(u)\}, \tag{2}$$

where $L(u)$ is a regularization function describing a prior knowledge on the image u. There exist an interest number of proposed regularizations functions in the literature, see for example [1–4]. The major idea of this approaches is to provide a global minimum to the optimization problem. In the same context, Maiseli et al. [5] proposed an appropriate regularization term defined as follows:

$$L(u) = \int_{\Omega} \frac{\frac{|\nabla u|}{\beta}\left(2 + \frac{|\nabla u|}{\beta}\right)}{1 + \left(\frac{|\nabla u|}{\beta}\right)^2} d\Omega. \tag{3}$$

with $\beta > 0$.

This process consist of combining the edge preserving effect of the Perona-Malik diffusion behavior [6] with the total variation [7] and the advantages soothing of backward diffusion model [8]. Which can be preserve the details and sharpen edges image. The evolution equation corespondent to the Euler-Lagrange of (2) with the regularization term (3) is given as

$$\frac{\partial u}{\partial t} = H^\top (Hu - \mathbf{u_0}) + \operatorname{div}\left(\frac{2 + \frac{|\nabla u|}{\beta}}{1 + \left(\frac{|\nabla u|}{\beta}\right)^2} \nabla u \right) \tag{4}$$

Since, if this equation preserve feature of the image, it suffers from the blurring effect in the flat regions. To overcome this problem, we present a anisotropic diffusion equation which increase the coherence-enhancing property and reduce blur. This process takes in consideration the advantages of Perona-Malik model in homogeneous regions [6], and the benefit of Weickert model effect [9] near edges and singularities. With Neumann boundary conditions the our equation is given as follows

$$\begin{cases} \frac{\partial u}{\partial t}(t,x) - \operatorname{div}\left(D\left(T_\rho\left(\nabla u_\sigma\right)\right)\nabla u\right) - H^\top (Hu - \mathbf{u_0}) = 0 \text{ on }]0, T[\times \Omega, \\ \left\langle D\left(T_\rho\left(\nabla u_\sigma\right)\right)\nabla u, n\right\rangle = 0 \text{ on }]0, T[\times \partial\Omega, \\ u(0,x) = u_0(x), \end{cases} \tag{5}$$

where D is an anisotropic diffusion matrix and T_ρ is the structure tensor given by

$$T_\rho(\nabla u_\sigma) = K_\rho * (\nabla u_\sigma \otimes \nabla u_\sigma) = K_\rho * (\nabla K_\sigma * u \nabla K_\sigma * u^\intercal). \tag{6}$$

where K_ρ and K_σ are two Gaussian convolution kernels such as $K_\tau(x) = \frac{1}{2\pi\tau^2}\exp(-\frac{|x|^2}{2\tau^2})$. The function D is chosen in order to preserve details image as much as possible.

The diffusion matrix D can be defined by using the eigenvectors and the eigenvalues of the tensor T_ρ as follows

$$D := h_+(\mu+, \mu_-)\theta_+\theta_+^{\mathsf{T}} + h_-(\mu_+, \mu_-)\theta_-\theta_-^{\mathsf{T}}, \tag{7}$$

with $\theta_{+/-}$ and $\mu_{+/-}$ are respectively the eigenvectors and the eigenvalues of the tensor T_ρ. The eigenvalues $\mu_{+/-}$ are computed as

$$\mu_{+/-} = \frac{1}{2}\left(trace(T_\rho) \pm \sqrt{trace^2(T_\rho) - 4\det(T_\rho)}\right). \tag{8}$$

This choice is depending on the geometric characteristics of the sharp image, see for example [10]. Indeed, if $\mu_{+/-} \approx 0$, the smoothing in flat regions is isotropic, while if $\mu_+ \gg \mu_- \approx 0$ (or $\mu_- \gg \mu_+ \approx 0$), the smoothing process is anisotropic and oriented along the straight sharp edges. However, if $\mu_+ \gg \mu_- \gg 0$, the smoothing takes into account singularities. Moreover, the functions $h_+(\mu_+, \mu_-)$ and $h_-(\mu_+, \mu_-)$ are defined neatly in order to satisfy the following smoothing proprieties:

1. Isotropic: the diffusion process is isotropic in the flat regions .
2. Anisotropic: the diffusion process is anisotropic near of sharp edges.

To define the functions $h_{+/-}$, many choices have been proposed [6, 9]. Here, we can consider the following functions, that allow to preserve more the image details.

$$\begin{cases} h_+(\mu_+, \mu_-) = \exp(-\frac{\mu_+}{k_1}), \\ h_-(\mu_+, \mu_-) = \exp(-\frac{\mu_-}{k_2})(1 - \exp(-\frac{\mu_+}{k_1})), \end{cases} \tag{9}$$

where k_1 and k_2 are two thresholds which define the diffusion along the eigenvectors θ_+ and θ_- respectively. Since, the desired diffusion depends according to the image details: edges or corner, the parameters k_1 and k_2 are used to define the sensitivity threshold that determine the variation in a given image. Using this proprieties, the Eq. (5) allows to reduce noise in flats regions and smooth contours. In addition, it preserves corners correctly.

New, in the following theorem we present the existence and uniqueness of a solution of our Eq. (5). We first give some interesting conditions:

1. The diffusion tensor D is infinitely continuous, i.e $D \in \mathcal{C}^\infty(\mathbb{R}^{2\times 2}, \mathbb{R}^{2\times 2})$.
2. D is a symmetric and positive-definite matrix.

In the following theorem, we prove the existence and uniqueness of solution to (5).

Theorem 11 *Given* $u_0 \in \mathbf{L}^\infty(\Omega)$*,* $\rho \geq 0$*,* $\sigma > 0$ *and* $T > 0$*. Under the above assumptions, the problem* (5) *admits a unique solution* u *in the space* $C((0, T); \mathbf{L}^2$

$(\Omega)) \cap \mathbf{L}^2((0,T);\mathbf{H}^1(\Omega))$. *Moreover, the solution depends continuously on* u_0 *with respect to* $\|.\|_{\mathbf{L}^2(\Omega)}$, *and* $u \in C((0,T)\times\overline{\Omega})$.

Proof See [11].

2 Discritization and Numerical Result

Now, we propose an appropriate discretizations of the multi-frame super resolution problem. The only difficulty is discretizing the diffusion term div($\mathbf{D}\nabla u$). In fact, there exist several schemes in the literature that allow to discritize this term. Here, we consider a classical technique which proposed by Weickert in [9].

Let us now denoted

$$\mathbf{D} = \begin{pmatrix} a & b \\ b & c \end{pmatrix}, \tag{10}$$

with

$$\begin{aligned} a &= \mu_1 \cos^2\alpha + \mu_2 \sin^2\alpha, \\ b &= (\mu_1 - \mu_2)\sin\alpha\cos\alpha, \\ c &= \mu_1 \sin^2\alpha + \mu_2\cos^2\alpha. \end{aligned}$$

where α is giving by the formula

$$\tan(2\alpha) = \frac{2T_{12}}{T_{11} - T_{22}},$$

with T_{12}, T_{11} and T_{22} are the components of the structure tensor T_ρ, that are

$$T_\rho(\nabla u_\sigma) = \begin{pmatrix} T_{11} & T_{12} \\ T_{12} & T_{22} \end{pmatrix}$$

So, the anisotropic diffusion term div($D\nabla u$) can be developed as

$$\begin{aligned} \operatorname{div}(D\nabla u) &= \operatorname{div}\left[\begin{pmatrix} a & b \\ b & c \end{pmatrix}\begin{pmatrix} \partial_x u \\ \partial_y u \end{pmatrix}\right] \\ &= \partial_x(a\partial_x u) + \partial_x\left(b\partial_y u\right) + \partial_y(b\partial_x u) + \partial_y\left(c\partial_y u\right). \end{aligned}$$

where

$$\begin{aligned} \partial_x(a\partial_x u) &= \partial_x a \partial_x u + a\partial_{xx} u \\ \partial_x\left(b\partial_y u\right) &= \partial_x b\partial_y u + b\partial_{xy} u \\ \partial_y(b\partial_x u) &= \partial_y b\partial_x u + b\partial_{yx} u, \\ \partial_y\left(c\partial_y u\right) &= \partial_y c\partial_y u + c\partial_{yy} u. \end{aligned}$$

Let as consider a discrete rectangular image domain with $N \times M$ pixels and denote by $u_{i,j}$, $i = 1, \dots N$ and $j = 1, \dots M$ this image. We also denote by $(\partial_x u)_{i,j}$, $(\partial_y u)_{i,j}$, $(\partial_{xx})_{i,j}$, $(\partial_{xy})_{i,j}$, $(\partial_{yy})_{i,j}$ the discrete version respectively of the operators $\partial_x u$, $\partial_y u$, ∂_{xx}, ∂_{xy}, ∂_{yy}, that are

$$(\partial_x u)_{i,j} = \begin{cases} u_{i+1,j} - u_{i,j} & \text{if } i < N \\ 0 & \text{if } i = N \end{cases}, \quad \left(\partial_y u\right)_{i,j} = \begin{cases} u_{i,j+1} - u_{i,j} & \text{if } j < M \\ 0 & \text{if } j = M \end{cases}. \tag{11}$$

The second derivatives operators are approximated as

$$(\partial_{xx})_{i,j} = \begin{cases} u_{i+1,j} - 2u_{i,j} + u_{i-1,j} & \text{si } 1 < i < N, \\ u_{i+1,j} - u_{i,j} & \text{si } i = 1, \\ u_{i-1,j} - u_{i,j} & \text{si } i = N \end{cases}$$

$$\left(\partial_{xy}\right)_{i,j} = \begin{cases} u_{i,j+1} - u_{i,j} - u_{i-1,j+1} + u_{i-,j} & \text{si } 1 < i < N, 1 < j < M \\ 0 & \text{si } i = 1, \\ 0 & \text{si } i = N \end{cases}$$

$$\left(\partial_{yx} u\right)_{i,j} = \begin{cases} u_{i+1,j} - u_{i,j} - u_{i+1,j-1} + u_{i,j-1} & \text{si } 1 < i < N, 1 < j < M \\ 0 & \text{si } j = 1 \\ 0 & \text{si } j = M \end{cases}$$

$$\left(\partial_{yy} u\right)_{i,j} = \begin{cases} u_{i,j+1} - 2u_{i,j} + u_{i,j-1} & \text{si } 1 < j < M \\ u_{i,j+1} - u_{i,j} & \text{si } j = 1, \\ u_{i,j-1} - u_{i,j} & \text{si } j = M \end{cases}$$

Using these notations, the discretization of the diffusion term can be written:

$$[\text{div}\,(\mathbf{D}\nabla u)]_{i,j} = (\partial_x a)_{i,j}(\partial_x u)_{i,j} + (a)_{i,j}(\partial_{xx} u)_{i,j} + (\partial_x b)_{i,j}(\partial_y u)_{i,j} \tag{12}$$

$$+ (b)_{i,j}(\partial_{xy} u)_{i,j} + (\partial_y b)_{i,j}(\partial_x u)_{i,j} + (b)_{i,j}(\partial_{yx} u)_{i,j} \tag{13}$$

$$+ (\partial_y c)_{i,j}(\partial_y u)_{i,j} + (c)_{i,j}(\partial_{yy} u)_{i,j}. \tag{14}$$

In fact, the standard discretization of (12) is equivalent to convolve the discrete image u at any point (i, j) by the Stencil matrix of size 3×3 (Table 1). Moreover, one of the conditions imposed by Weickert [9] on the matrix associated to the diffusion term is the positivity of its elements, which is required for the stability of this schema.

If we denote by $A_{i,j} * u_{i,j}$ the discretization of div $(\mathbf{D}\nabla u)$ and S the Stencil matrix giving as:

$$S = \begin{pmatrix} s_{11} & s_{12} & s_{13} \\ s_{21} & s_{22} & s_{23} \\ s_{31} & s_{32} & s_{33} \end{pmatrix}.$$

Hence, $A_{i,j} * u_{i,j}$ is calculated by:

Table 1 Non negative stencil discretization

$\frac{\lvert b_{i-1,j+1}\rvert - b_{i-1,j+1}}{4} + \frac{\lvert b_{i,j}\rvert - b_{i,j}}{4}$	$\frac{c_{i,j+1}+c_{i,j}}{2} - \frac{\lvert b_{i,j+1}\rvert + \lvert b_{i,j}\rvert}{2}$	$\frac{\lvert b_{i+1,j+1}\rvert + b_{i+1,j+1}}{4} + \frac{\lvert b_{i,j}\rvert + b_{i,j}}{4}$
$\frac{a_{i-1,j}+a_{i,j}}{2} - \frac{\lvert b_{i-1,j}\rvert + \lvert b_{i,j}\rvert}{2}$	$-\frac{a_{i-1,j}+2a_{i,j}+a_{i+1,j}}{2} - \frac{\lvert b_{i-1,j+1}\rvert - b_{i-1,j+1} + \lvert b_{i+1,j+1}\rvert + b_{i+1,j+1}}{4} - \frac{\lvert b_{i-1,j-1}\rvert + b_{i-1,j-1} + \lvert b_{i+1,j-1}\rvert - b_{i+1,j-1}}{4} + \frac{\lvert b_{i-1,j}\rvert + \lvert b_{i+1,j}\rvert + \lvert b_{i,j-1}\rvert + \lvert b_{i,j+1}\rvert + 2\lvert b_{i,j}\rvert}{2} - \frac{c_{i,j-1}+2c_{i,j}+c_{i,j+1}}{2}$	$\frac{a_{i+1,j}+a_{i,j}}{2} - \frac{\lvert b_{i+1,j}\rvert + \lvert b_{i,j}\rvert}{2}$
$\frac{\lvert b_{i-1,j-1}\rvert + b_{i-1,j-1}}{4} + \frac{\lvert b_{i,j}\rvert + b_{i,j}}{4}$	$\frac{c_{i,j-1}+c_{i,j}}{2} - \frac{\lvert b_{i,j-1}\rvert + \lvert b_{i,j}\rvert}{2}$	$\frac{\lvert b_{i+1,j-1}\rvert - b_{i+1,j-1}}{4} + \frac{\lvert b_{i,j}\rvert - b_{i,j}}{4}$

$$\begin{aligned} A_{i,j} * u_{i,j} = {} & u_{i-1,j-1}s_{11} + u_{i-1,j}s_{12} + u_{i-1,j+1}s_{13} \\ & + u_{i,j-1}s_{21} + u_{i,j}s_{22} + u_{i,j+1}s_{23} \\ & + u_{i+1,j-1}s_{31} + u_{i+1,j}s_{32} + u_{i+1,j+1}s_{33}. \end{aligned}$$

Using a forward difference approximation, we discretize the time derivative $\frac{\partial u_{i,j}}{\partial t}$ by $\frac{u_{i,j}^{n+1} - u_{i,j}^{n}}{dt}$.

Finally, the discretization of the Eq. (5) can be rewritten as:

$$\frac{u_{i,j}^{n+1} - u_{i,j}^{n}}{dt} = A_{i,j}^{n} * u_{i,j}^{n} \quad \Leftrightarrow \quad u_{i,j}^{n+1} = \left(I + dt A_{i,j}^{n}\right) * u_{i,j}^{n}$$

Now, we present some numerical simulation and also real results for testing the performance of the present multi-frame SR method. Comparison with some competitive and available SR approach is carried out. Namely, TV [12], BTV [1], TV+BTV [3], and model of nonlinear diffusion regularizer [5]. The presented super-resolution equation is tested on a large and reasonable benchmark image, we give only three of them. In the first simulated test, we degraded each of the real HR images of the Castle, EIAcen and Penguin to create the corresponding set of thirty synthetic lower-resolution images. This set of images is computed from the original images such as each frame is degraded by blur of Gaussian filter with a 3×3 and standard deviation of 1.5. So, the blurred sequence image are down-sampled horizontally and vertically by a factor of $r = 4$ and finally we add a Gaussian noise, to down-sampled frames, with $\sigma = 20$. Then, we tested different multi-frame super-resolution model selected above, and the proposed model to recover the respective original images of the corrupted ones. To evaluate and compare the robustness of the our model with the others, a quantitative comparison is needed, we have used two measures: peak-signal-to-noise ratio (PSNR) [13], and mean structure similarity (SSIM) [14]. The PSNR metrics signal strength relative to noise in the image and is given by

$$PSNR = 10 \log_{10}\left(\frac{255^2}{mse}\right),$$

with the mse is the mean squared error defined as

$$mse = \frac{1}{MN}\Sigma_{i=1}^{M}\Sigma_{j=1}^{N}(Y(i,j) - X(i,j))^2.$$

The $SSIM$ is calculated on multiple windows of given image, i.e. the measurement between two windows x and y of size $N \times N$ is defined by

$$SSIM(x,y) = \frac{(2\mu_x\mu_y + c_1)(2\sigma_x\sigma_y + c_2)(2\,cov_{xy} + c_3)}{(\mu_x^2 + \mu_y^2 + c_1)(\sigma_x^2 + \sigma_y^2 + c_2)(\sigma_x\sigma_y + c_3)},$$

with the variables, are respectively, given for x and y as follows: μ_x and μ_y, mean; σ_x^2 and σ_y^2, variance; cov_{xy}, covariance; $c_1 = (k_1 l)^2$, $c_2 = (k_2 L)^2$ are two stabilizing constants; and l the dynamics of the pixel values, 255 for 8-bit encoded image. This measure gives an indication on the quality of the image based on the known characteristics of human visual system.

The results presented in Figs. 1, 2 and 3 prove that, visually, the our method outperforms the others. In addition, the results given in the Tables 2 and 3 confirm that the our method is usually with the higher value of PSNR and SSIM. Note that we take the optimal parameters according to the best PSNR and SSIM values in all the tests for the other models. Characteristically, the execution of the main implementation on MATLAB R2014 requires on the average about $7 \sim 14$ s on a i7 3.4 GHz Quad-Core computer for 256×256 images. Furthermore, as is shown in Table 4, quantitative results prove that the proposed method, usually, is robust in reconstruction with respect to the PSNR values. Despite the presented results, our method suffers from the weakness that it does not preserve more the texture.

Finally, we can summaries that the presented results prove that the studied model reconstruct a sharper image with less blur compared with the other models.

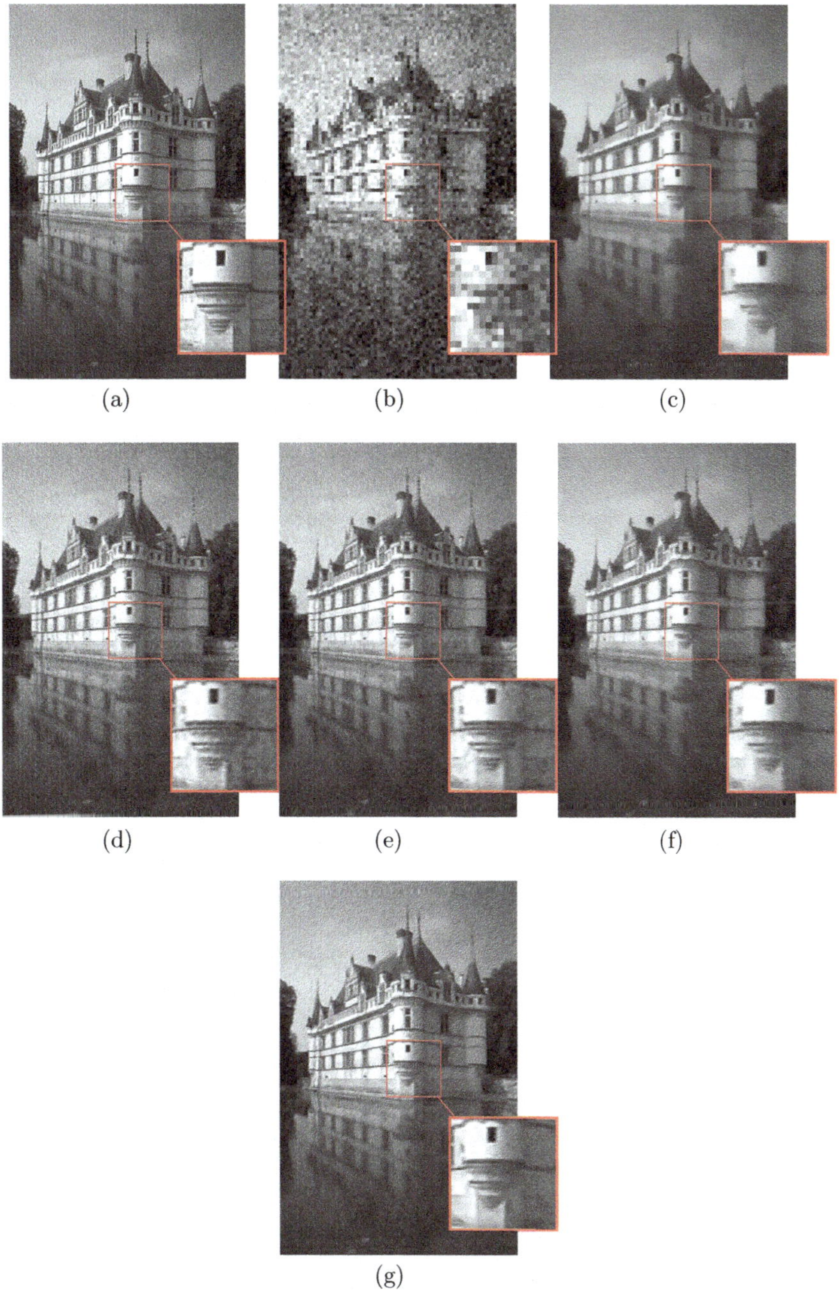

Fig. 1 The results obtained by applying different methods to LR (*Castle* sequence) **a** HR image, **b** One LR image, **c** Farsiu et al. [1], **d** Marquina et al. [12], **e** Maiseli et al. [5], **f** Laghrib et al. [3], **g** our method

Table 2 The PSNR table

Image	PSNR			
	Marquina et al. [12]	Maiseli et al. [5]	Laghrib et al. [3]	Our method
Champignon	29.26	29.13	29.32	**30.02**
Penguin	31.58	32.48	32.22	**32.98**
Castle	27.03	27.12	26.45	**27.30**
EIAcen	20.56	21.24	20.81	**22.10**
Lena	29.16	29.79	29.23	**30.21**
Bird	33.93	34.70	34.78	**35.12**
Peppers	30.67	**31.45**	31.18	31.11
Goldhill	29.36	30.22	29.92	**30.42**
Lake	30.89	31.16	31.62	**31.92**
Baboon	27.50	27.51	27.50	**27.93**
Average	28.99	29.34	29.33	**29.99**

Table 3 The SSIM table

Image	SSIM			
	Marquina et al. [12]	Maiseli et al. [5]	Laghrib et al. [3]	Our method
Champignon	0.815	0.822	0.800	**0.838**
Penguin	0.800	0.882	0.881	**0.886**
Castle	0.744	0.796	**0.811**	0.800
EIAcen	0.823	0.870	0.751	**0.878**
Lena	0.821	0.840	0.843	**0.855**
Bird	0.843	0.873	0.877	**0.898**
Peppers	0.756	0.770	0.759	**0.779**
Goldhill	0.831	0.867	0.865	**0.874**
Lake	0.816	0.821	0.821	**0.832**
Baboon	0.712	0.741	0.743	**0.749**
Average	0.821	0.823	0.804	**0.849**

Table 4 The PSNR table of the reconstructed Fish and Barbara images using different σ noise values

Image	Method	$\sigma = 10$	$\sigma = 30$	$\sigma = 50$
Fish	Maiseli et al. [5]	34.79	30.98	28.74
	Laghrib et al. [3]	34.12	30.87	29.12
	Our method	**34.95**	**31.70**	**30.47**
Barbara	Maiseli et al. [5]	**29.13**	**25.86**	24.03
	Laghrib et al. [3]	28.38	25.16	23.91
	Our method	28.94	25.84	**24.54**

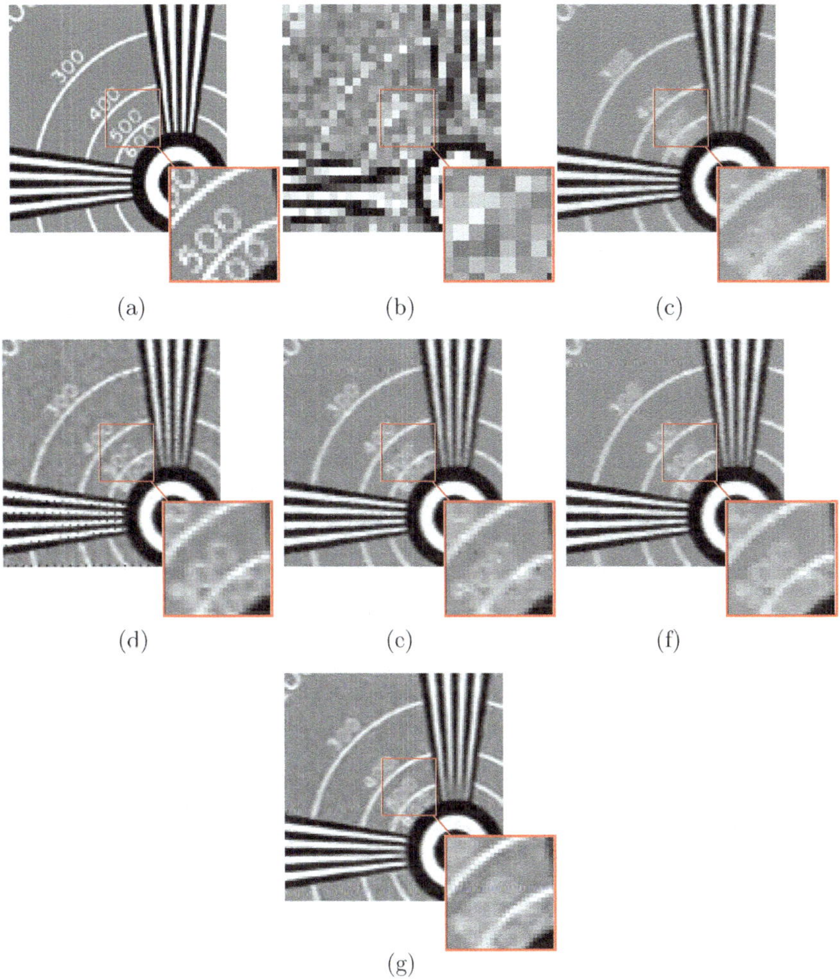

Fig. 2 The results obtained by applying different SR methods to LR *EIAcen* sequence **a** HR image, **b** One LR image, **c** Farsiu et al. [1], **d** Marquina et al. [12], **e** Maiseli et al. [5], **f** Laghrib et al. [3], **g** our method

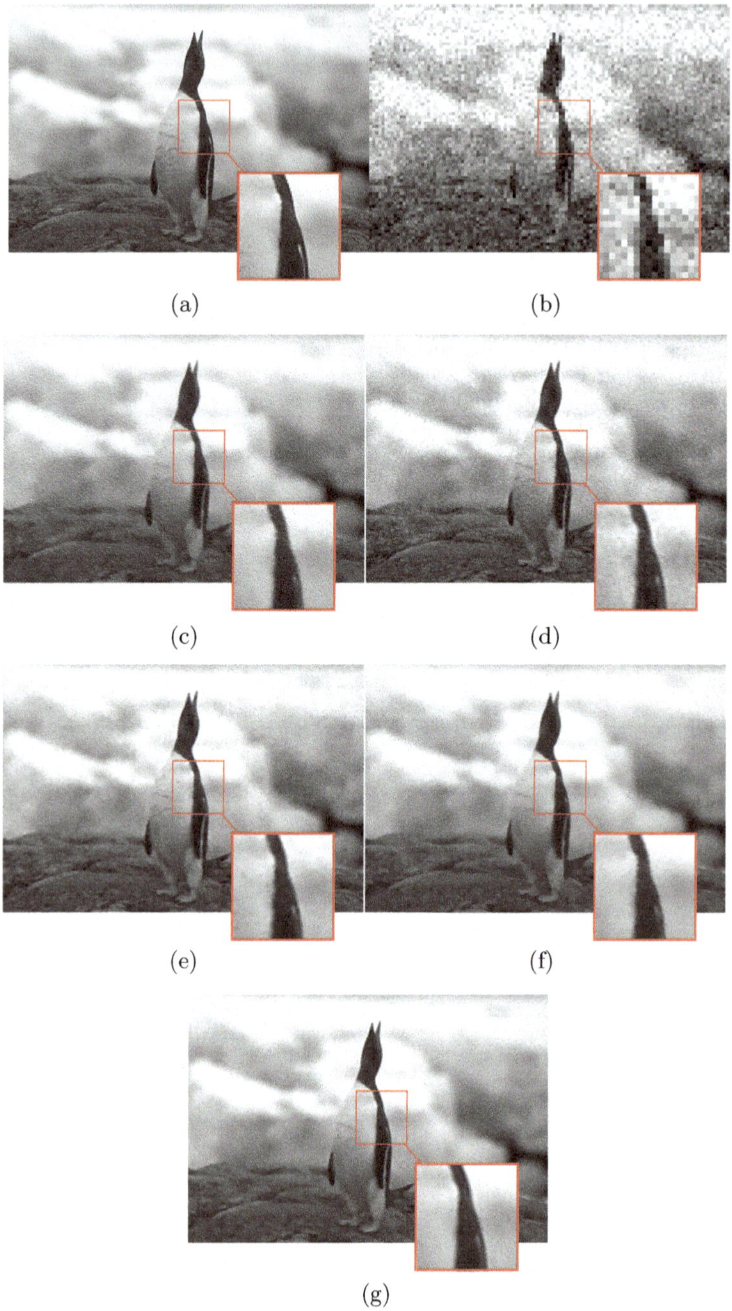

Fig. 3 The results obtained by applying different methods to (*Penguin* sequence) **a** HR image, **b** One LR image, **c** Farsiu et al. [1], **d** Marquina et al. [12], **e** Maiseli et al. [5], **f** Laghrib et al. [3], **g** our method

References

1. S. Farsiu, D. Robinson, M. Elad, P. Milanfar, Advances and challenges in super-resolution. Int. J. Imaging Syst. Technol. **14**(2), 47–57 (2004)
2. X. Zeng, L. Yang, A robust multiframe super-resolution algorithm based on half-quadratic estimation with modified BTV regularization. Digit. Signal Process. **23**(1), 98–109 (2013)
3. A. Laghrib, A. Hakim, S. Raghay, A combined total variation and bilateral filter approach for image robust super resolution. EURASIP J. Image Video Process. **2015**(1), 1–10 (2015)
4. S. Zhao, H. Liang, M. Sarem, A generalized detail-preserving super-resolution method. Signal Process. **120**, 156–173 (2016)
5. B.J. Maiseli, N. Ally, H. Gao, A noise-suppressing and edge-preserving multiframe super-resolution image reconstruction method. Signal Process. Image Commun. **34**, 1–13 (2015)
6. P. Perona, J. Malik, Scale-space and edge detection using anisotropic diffusion. IEEE Trans. Pattern Anal. Mach. Intell. **12**(7), 629–639 (1990)
7. L.I. Rudin, S. Osher, E. Fatemi, Nonlinear total variation based noise removal algorithms. Phys. D Nonlinear Phenomena **60**(1), 259–268 (1992)
8. Y.L. You, W. Xu, A. Tannenbaum, M. Kaveh, Behavioral analysis of anisotropic diffusion in image processing. IEEE Trans. Image Process. **5**(11), 1539–1553 (1996)
9. J. Weickert, An isotropic diffusion in image processing. Teubner Stuttgart (1998)
10. J. Weickert, Coherence-enhancing diffusion of colour images. Image Vis. Comput. **17**(3), 201–212 (1999)
11. I. El Mourabit, M. El Rhabi, A. Hakim, A. Laghrib, E. Moreau, A new denoising model for multi-frame super-resolution image reconstruction. Signal Process. **132**, 51–65 (2017)
12. A. Marquina, S.J. Osher, Image super-resolution by TV-regularization and Bregman iteration. J. Sci. Comput. **37**(3), 367–382 (2008)
13. Z. Wang, A.C. Bovik, Mean squared error: love it or leave it? A new look at signal fidelity measures. IEEE Signal Process. Mag. **26**(1), 98–117 (2009)
14. Z. Wang, A.C. Bovik, H.R. Sheikh, E.P. Simoncelli, Image quality assessment: from error visibility to structural similarity. IEEE Trans. Image Process. **13**(4), 600–612 (2004)

Acceleration of the KMF Algorithm Convergence to Solve the Cauchy Problem for Poisson's Equation

Abdeljalil Nachaoui, Fatima Aboud, and Mourad Nachaoui

Abstract In this article we propose a new demonstration of the convergence of the relaxed iterative JN algorithm for solving the inverse Cauchy problem. We give in particular the convergence interval as well as the interval of the convergence acceleration of KMF algorithm.

Keywords Inverse Cauchy problem · Data completion · Missing boundary condition · Relaxed iterative algorithm · Convergence acceleration · Separation method

1 Introduction

In this work, we are interested in solving an inverse Cauchy problem for Poisson equation, which consists in the reconstruction of the data on a non accessible part of the boundary based on the measurements available on an other part. There are many approaches to solve this ill-posed problem [1–9]. One of them is the KMF alternating algorithm proposed in [6]. This approach is based on reducing the ill-posed Cauchy problem to a sequence of mixed well-posed boundary value problems, with Dirichlet and/or Neumann conditions on the non-accessible part. The authors theoretically show that the algorithm produces a convergent sequence whose limit is the solution of the Cauchy problem, but they give no numerical application. The first application of this algorithm is made by Jourhman and Nachaoui in [10] where they have shown that the algorithm was found to produce an accurate and stable numerical solution for the Cauchy problem. However, they pointed out a major drawback of the KMF algorithm,

A. Nachaoui (✉)
Laboratoire de Mathématiques Jean Leray, Université de Nantes, Nantes, France

F. Aboud
Department of Mathematics, College of Science, University of Diyala, Baqubah, Iraq

M. Nachaoui
Laboratoire de Mathématiques et Applications, Université Sultan Moulay Slimane, Béni-Mellal, Maroc

A. Nachaoui et al. (eds.), *Mathematical Control and Numerical Applications*,
Springer Proceedings in Mathematics & Statistics 372,
https://doi.org/10.1007/978-3-030-83442-5_4

namely its slowness (a large number of iterations is necessary to reach convergence) in particular if the initial guess is far from the exact solution. They proposed to remedy this drawback by a relaxation of the given Dirichlet data. In [11] Jourhmane-Nachaoui showed numerically that, their JN relaxed algorithm not only does it keep the same advantages as the classic KMF algorithm but in addition this procedure drastically reduced the number of iterations required to achieve convergence for the considered Cauchy problems for some relaxation parameters θ chosen in an interval $(1, \theta_1)$ with $\theta_1 < 2$. In 2002 Jourhmane and Nachaoui gave a theoretical result of the convergence of their relaxed algorithm based on Krasnolskii's fixed point theory [12]. They proved that there is a real number θ_2 such that the algorithm converges to the exact solution, provided that relaxation parameter θ belongs to $(0, \theta_2)$ with $\theta_2 \leq 2$. However, the numerical tests performed revealed that these algorithms are convergent for a larger interval $(0, \theta_2)$. The numerical tests performed showed that within the interval of convergence $(0, \theta_2)$ there are large variations in the rate of convergence of the algorithms. Since these works, several authors have applied or have developed relaxed algorithms for the resolution of various Cauchy problems, we cite for example the works of Ellabib and Nachaoui (2008) [5] and Marin and Johansson (2010)[7] an others [13–16]. However until now no estimates are available for the limits of the convergence and the acceleration of convergence intervals $(0, \theta_2)$ and $(1, \theta_1)$. Therefore it is the purpose of this paper to revisit the convergence results of the relaxed JN algorithm [12] in order to find expressions for these two parameters.

In what follows, we use the theory of separation of variables to develop a simple and understandable demonstration of convergence. We also prove theoretically a result of convergence acceleration which, until now, was just investigated numerically. We give in particular an estimate of the parameters θ_1 and θ_2 according to the data of the Cauchy problem considered.

2 Cauchy Problem for Poisson's Equation and JN Relaxed Algorithm

Let Ω be a bounded domain in R^d with the Lipschitz boundary Γ. Let Γ be divided into four disjoint parts Γ_0, Γ_1, Γ_2 and Γ_3. Denote by ν the outward unit normal to the boundary Γ and consider the following Cauchy problem for the Poisson's equation:

$$\Delta u = f \ \text{ in } \Omega, \quad u|_{\Gamma_1} = f_1, \quad \partial_\nu u|_{\Gamma_1} = g_1, \quad u|_{\Gamma_2} = f_2, \quad u|_{\Gamma_3} = f_3. \tag{1}$$

2.1 *The JN Relaxed Algorithm*

The JN relaxation algorithm of Jourhmane-Nachaoui [10, 11], that we consider to solve the Cauchy problem (1) is described as follows:

Step 1. Choose an initial approximation $\phi^{(0)}$ on the boundary (Γ_0), and solve the well-posed mixed boundary value problem

$$\Delta u^{(0)} = f \ \text{ in } \Omega, \ \ \partial_\nu u|_{\Gamma_1}^{(0)} = g_1, \ \ u|_{\Gamma_2}^{(0)} = f_2, \ \ u|_{\Gamma_3}^{(0)} = f_3, \ \ u|_{\Gamma_0}^{(0)} = \phi^{(0)} \quad (2)$$

Step 2. Solve the well-posed mixed boundary value problem for $n = 1, 2, 3, \ldots$

$$\Delta u^{(2n+1)} = f \text{ in } \Omega, \ u|_{\Gamma_1}^{(2n+1)} = f_1, \ u|_{\Gamma_2}^{(2n+1)} = f_2, \ u|_{\Gamma_3}^{(2n+1)} = f_3, \ \partial_\nu u|_{\Gamma_0}^{(2n+1)} = \partial_\nu u|_{\Gamma_0}^{(2n)} \quad (3)$$

Step 3. Solve the well-posed mixed boundary value problem for $n = 1, 2, 3, \ldots$

$$\Delta u^{(2n)} = f \ \text{ in } \Omega, \ \ \partial_\nu u|_{\Gamma_1}^{(2n)} = g_1, \ \ u|_{\Gamma_2}^{(2n)} = f_2, \ \ u|_{\Gamma_3}^{(2n)} = f_3, \ \ u|_{\Gamma_0}^{(2n)} = \phi|_{\Gamma_0}^{(n)} \quad (4)$$

$$\phi^{(n)} = \theta u^{(2n-1)} + (1-\theta)\phi^{(n-1)}, \quad for \ \ n \geq 1 \ \ \theta \in (0, 2]. \quad (5)$$

Step 4. Repeat step 2 and step 3 consequently untill a commended stopping criterion is achieved.

Note that for $\theta = 1$, we get the standard algorithm of Kozlov-Mas'ya [6].

The main result of this work is summarized in the following theorem:

Theorem 1 *Considering the Poisson problem defined in (2)–(4). We have: For all* $\theta \in \left(0, \frac{2}{1-\left(\tanh(\frac{\pi}{a}b)\right)^2}\right)$ *the sequence* $\{u^{(n)}\}$ *converges to the solution of the Cauchy problem in (1) independently of the initial value* $\phi^{(0)}$.

To prove the convergence of this last algorithm we recall the following lemma in which Jourhmane-Nachaoui [12] give the idea of proving the convergence of the algorithm to the solution in the domain Ω by proving its convergence on the boundary Γ_0 (for the proof see [12]):

Lemma 1 *[12] If the sequence* $u|_{\Gamma_0}^{(2n)}$ *converges in* $\mathcal{L}^2(\Gamma_0)$, *then the sequence* $u|_{\Gamma_0}^{(n)}$ *converges in* $\mathcal{H}^1(\Omega)$ *to the solution of (1), where* $\mathcal{H}^1(\Omega)$ *is the Sobolev space and* $\mathcal{L}^2(\Gamma)$ *is the space of square integrable functions on* Γ.

3 Convergence Analysis

Note that, using (5) we can rewrite the non-homogeneous condition of (4) in the following form:

$$\begin{aligned} u^{(2n)} &= \theta u^{(2n-1)} + (1-\theta)\phi^{(n-1)} \\ &= \theta u^{(2n-1)} + (1-\theta)u^{(2(n-1))} \end{aligned} \quad (6)$$

Let $w^{(n)} = u - u^{(n)}$, then from the JN relaxed algorithm, $w^{(n)}$ is given as follows:

For $n = 0$, $w^{(0)}$ is the solution of the boundary value problem:

$$\Delta w^{(0)} = 0 \text{ in } \Omega, \quad \partial_\nu w^{(0)}|_{\Gamma_1} = 0 \quad w^{(0)}|_{\Gamma_2} = 0 \quad w^{(0)}|_{\Gamma_3} = 0 \quad w^{(0)}|_{\Gamma_0} = u|_{\Gamma_0} - \phi^{(0)} \tag{7}$$

For $n = 0, 2, 3, \ldots$ $w^{(2n+1)}$ is the solution of the (odd) problem:

$$\Delta w^{(2n+1)} = 0 \text{ in } \Omega, \ w^{(2n+1)}|_{\Gamma_1} = 0, \ w^{(2n+1)}|_{\Gamma_2} = 0, \ w^{(2n+1)}|_{\Gamma_3} = 0, \ \partial_\nu w^{(2n+1)}|_{\Gamma_0} = \partial_\nu w^{(2n)}|_{\Gamma_0} \tag{8}$$

For $n = 1, 2, 3, \ldots$ $w^{(2n)}$ is the solution of the (even) problem:

$$\Delta w^{(2n)} = 0 \ \text{ in } \Omega, \quad \partial_\nu w^{(2n)}|_{\Gamma_1} = 0, \quad w^{(2n)}|_{\Gamma_2} = 0, \quad w^{(2n)}|_{\Gamma_3} = 0, \quad w^{(2n)}|_{\Gamma_0} = v^{(n)} \tag{9}$$

where $v^{(n)} = u|_{\Gamma_0} - \phi^{(n)}$, with $\phi^{(n)}$ defined in (5). Using (6) and the fact that $u^{(2n)}|_{\Gamma_0} = \phi^{(n)}|_{\Gamma_0}$, we obtain

$$v^{(n)} = \theta w^{(2n-1)} + (1-\theta) w^{2(n-1)}, \quad for \ \ n \geq 1 \tag{10}$$

Remark 1 Note that the study of the convergence of $u^{(n)}$ to u, the solution of 1, is equivalent to show that $w^{(n)}$ converges to 0.

Based on this remark, our study of the convergence will be on $w^{(n)}$. To illustrate our technique and without lose of generality let Ω be the rectangular domain defined by $\Omega = (0, a) \times (0, b)$ with a and b are non negatives reals constantes. Let $\Gamma_1 = (0, a) \times \{0\}$, $\Gamma_2 = \{a\} \times (0, b)$, $\Gamma_3 = \{0\} \times (0, b)$ and $\Gamma_0 = (0, a) \times \{b\}$. The method of resolution of problems (7)–(9) by separation of the variables consists in looking for solutions of the form (we separate the variables x and y):

$$w^{(n)} = X^{(n)}(x) Y^{(n)}(y)$$

By replacing this expression in the Laplace equation of (7)–(9), we have:

$$\frac{d^2 X^{(n)}}{dx^2} Y^{(n)} + \frac{d^2 Y^{(n)}}{dy^2} X^{(n)} = 0 \tag{11}$$

Now, if we divide this expression by the product $X^{(n)} Y^{(n)}$, we get:

$$\frac{1}{X^{(n)}} \frac{d^2 X^{(n)}}{dx^2} = -\frac{1}{Y^{(n)}} \frac{d^2 Y^{(n)}}{dy^2} \tag{12}$$

The equality requires that both ratios to be constant, it is the only possibility for functions of different variables. We then introduce the separation constant $\lambda = \alpha^2$ which must be defined. So we have

$$\frac{1}{X^{(n)}}\frac{d^2X^{(n)}}{dx^2} = -\alpha^2 \ and \ \frac{1}{Y^{(n)}}\frac{d^2Y^{(n)}}{dy^2} = \alpha^2 \tag{13}$$

We are then brought to solve the following two problems for $X^{(n)}$ and $Y^{(n)}$,:

$$\begin{cases} d_x^2X^{(2n)} + \alpha^2 X^{(2n)} = 0 & X^{(2n)}(0) = X^{(2n)}(a) = 0 \\ d_y^2Y^{(2n)} - \alpha^2 Y^{(2n)} = 0 & d_yY^{(2n)}(0) = 0 \end{cases} \tag{14}$$

and

$$\begin{cases} d_x^2X^{(2n+1)} + \alpha^2 X^{(2n+1)} = 0 & X^{(2n+1)}(0) = X^{(2n+1)}(a) = 0 \\ d_y^2Y^{(2n+1)} - \alpha^2 Y^{(2n+1)} = 0 & Y^{(2n+1)}(0) = 0 \end{cases} \tag{15}$$

From the X-problem in (14) we get

$$X^{(2n)} = \big(A\cos(\alpha x) + B\sin(\alpha x). \tag{16}$$

Condition $X^{(2n)}(0) = 0$ imply that $A = 0$, so we obtain

$$X^{(2n)} = B\sin(\alpha x). \tag{17}$$

Condition $X^{(2n)}(a) = 0$ gives $B\sin(\alpha a) = 0$ which impose that (apart from $B = 0$, which is a trivial solution): $\alpha = \frac{m\pi}{a}$, with $n = 1, 2, 3, \ldots$.

Consequently, $\lambda_n = (\frac{m\pi}{a})^2$ are the eigenvalues of the X-problem. The eigenfunctions associated with the eigenvalues λ_n are written:

$$X_m^{(2n)} = B_m \sin(\frac{m\pi}{a}x). \tag{18}$$

Now the Y-problem in (14) imply that

$$Y_m^{(2n)} = \big(C_m\cosh(\frac{m\pi}{a}y) + D_m\sinh(\frac{m\pi}{a}y)\big). \tag{19}$$

Using the boundary condition $d_yY_m^{(2n)}(0) = 0$, we get that $D_m = 0$ and so

$$Y_m^{(2n)} = C_m\cosh(\frac{m\pi}{a}y). \tag{20}$$

The general solution $w^{(2n)}$ of the even problem is the superposition of the set of solutions (we must consider all the eigenfunctions).

$$w^{(2n)}(x, y) = \sum_{m=1}^{\infty} A_m^{(2n)} \sin\big(\frac{m\pi}{a}x\big)\cosh\big(\frac{m\pi}{a}y\big). \tag{21}$$

We use the same technique for (15) to obtain that the solution for the odd problem is given by

$$w^{(2n+1)}(x,y)=\sum_{m=1}^{\infty}A_m^{(2n+1)}\sin\left(\frac{m\pi}{a}x\right)\sinh\left(\frac{m\pi}{a}y\right). \tag{22}$$

To obtain a solution of problems (7)–(9), the use of conditions on Γ_0 allows to find the expression of $A_m^{(2n)}$ and $A_m^{(2n+1)}$. This is illustrated in the following lemma:

Lemma 2 *Considering $w^{(2n)}$ and $w^{(2n+1)}$ in the problems (7)–(9), then the coefficients $A_m^{(2n)}$ and $A_m^{(2n+1)}$ in (21) and (22) are given by*

$$A_m^{(2n)}=A_m^{(0)}\Big[\theta\big(\tanh(\beta_m b)\big)^2+(1-\theta)\Big]^n \tag{23}$$

$$A_m^{(2n+1)}=A_m^{(0)}\Big[\theta\big(\tanh(\beta_m b)\big)^2+(1-\theta)\Big]^n\tanh(\beta_m b) \tag{24}$$

where $\beta_m=\frac{m\pi}{a}$ and

$$A_m^{(0)}=\frac{2}{a\cosh(\beta_m b)}\int_0^a\big(u(x,b)-\phi^{(0)}\big)\sin(\beta_m x)dx. \tag{25}$$

Proof We prove the lemma by induction on n. For $n=0$, we apply the non-homogeneous condition of the problem (7) $w|_{\Gamma_0}^{(0)}=u|_{\Gamma_0}-\phi^{(0)}$ to get

$$\sum_{m=1}^{\infty}A_m^{(0)}\sin\big(\beta_m x\big)\cosh\big(\beta_m b\big)=u|_{\Gamma_0}-\phi^{(0)},\quad \text{for } x\in(0,a). \tag{26}$$

Note that functions $\{\sin(\frac{n\pi}{a}x)\}_{n=1}^{\infty}$ are orthogonal and satisfies

$$\int_0^a\sin(\frac{l\pi}{a}x)\sin(\frac{m\pi}{a}x)dx=\frac{a}{2}\delta_{lm}, \tag{27}$$

where δ_{lm} is the Kronecker delta. Thus, multiply both side of (26) by $\sin(\beta_k x)$, integrate with respect to x and the orthogonality (27) leads to

$$A_k^{(0)}\frac{a}{2}\cosh\big(\beta_k b\big)=\int_0^a(u|_{\Gamma_0}-\phi^{(0)})\sin(\beta_k x)dx,$$

which gives formula (25).

The coefficient $A_m^{(1)}$ is obtained using the non-homogeneous boundary conditions of problems (8)

$$\partial_\nu w^{(1)}|_{\Gamma_0}=\partial_\nu w^{(0)}|_{\Gamma_0}$$

written as

$$\sum_{m=1}^{\infty} A_m^{(1)} \beta_m \sin\left(\beta_m x\right) \cosh\left(\beta_m b\right) = \sum_{m=1}^{\infty} A_m^{(0)} \beta_m \sin\left(\beta_m x\right) \sinh\left(\beta_m b\right).$$

Multiplying by $\sin\left(\beta_k x\right)$ and integrating with respect to x, we get

$$A_k^{(1)} \frac{a}{2} (\beta_k) \cosh\left(\beta_k b\right) = A_k^{(0)} \frac{a}{2} (\beta_k) \sinh\left(\beta_k b\right), \quad \text{for } k = 1, 2, \ldots$$

so

$$A_m^{(1)} - A_m^{(0)} \tanh\left(\beta_m b\right), \quad \text{for } m = 1, 2, \ldots \tag{28}$$

which proves formula (24) for $n = 0$.

Now, for $n = 1$, we find $A_m^{(2)}$ and $A_m^{(3)}$ using the non-homogeneous boundary conditions of problems (9) and (8) respectively.
From $w|_{\Gamma_0}^{(2)} = \theta w^{(1)}(x, b) + (1 - \theta) w^{(0)}(x, b)$ we obtain

$$\sum_{m=1}^{\infty} A_m^{(2)} \sin\left(\beta_m x\right) \cosh\left(\beta_m b\right) = \theta \sum_{m=1}^{\infty} A_m^{(0)} \tanh\left(\beta_m b\right) \sin\left(\beta_m x\right) \sinh\left(\beta_m b\right)$$
$$+(1 - \theta) \sum_{m=1}^{\infty} A_m^{(0)} \sin\left(\beta_m x\right) \cosh\left(\beta_m b\right).$$

Again, multiplying the above equality by $\sin(\beta_k x)$, integrating with respect to x and using the orthogonality (27), we get

$$A_k^{(2)} \frac{a}{2} \cosh\left(\beta_k b\right) = \theta A_k^{(0)} \tanh\left(\beta_k b\right) \frac{a}{2} \sinh\left(\beta_k b\right) + (1 - \theta) \frac{a}{2} \cosh\left(\beta_k b\right),$$

which gives

$$A_k^{(2)} = A_k^{(0)} \left[\theta \left(\tanh\left(\beta_k b\right)\right)^2 + (1 - \theta)\right], \quad \text{for } k = 1, 2, \ldots$$

This proves formula (23) for $n = 1$.
Now, $A_m^{(3)}$ is obtained in the same way as $A_m^{(1)}$ and this gives

$$A_m^{(3)} = A_m^{(2)} \tanh\left(\beta_m b\right) = A_m^{(0)} \left[\theta \left(\tanh\left(\beta_m b\right)\right)^2 + (1 - \theta)\right] \tanh\left(\beta_m b\right). \tag{29}$$

Finally, for $n > 1$ we suppose that these formulas are satisfied for $n = k$ and we show that they remain verified for $n = k + 1$.

From problem (9), we have for $n = k + 1$

$$w^{(2(k+1))}(x,y)=\sum_{m=1}^{\infty}A_m^{(2(k+1))}\sin\left(\beta_m x\right)\cosh\left(\beta_m y\right). \tag{30}$$

So, using the non-homogeneous boundary condition

$$\begin{aligned} w^{(2(k+1))}|_{\Gamma_0} &= \theta w^{(2(k+1)-1)}|_{\Gamma_0}+(1-\theta)w^{(2((k+1)-1))}|_{\Gamma_0}\\ &= \theta w^{(2k+1))}|_{\Gamma_0}+(1-\theta)w^{(2k)}|_{\Gamma_0}, \end{aligned}$$

thus

$$\begin{aligned} \sum_{m=1}^{\infty}A_m^{(2(k+1))}\sin\left(\beta_m x\right)\cosh\left(\beta_m b\right) = \theta\sum_{m=1}^{\infty}A_m^{(0)}&\left[\theta\left(\tanh(\beta_m b)\right)^2+(1-\theta)\right]^k \tanh(\beta_m b)\\ &\sin\left(\beta_m x\right)\sinh\left(\beta_m b\right)\\ &+(1-\theta)\sum_{m=1}^{\infty}A_m^{(0)}\left[\theta\left(\tanh(\beta_m b)\right)^2+(1-\theta)\right]^k\\ &\sin\left(\beta_m x\right)\cosh\left(\beta_m b\right). \end{aligned}$$

Again, multiplying the above equality by $\sin(\beta_l x)$, integrating and using the argument of orthogonality, we get

$$\begin{aligned} A_l^{(2(k+1))} &= \theta A_l^{(0)}\left[\theta\left(\tanh(\beta_l b)\right)^2+(1-\theta)\right]^k\left[\tanh(\beta_l b)\right]^2\\ &\quad+(1-\theta)A_l^{(0)}\left[\theta\left(\tanh(\beta_l b)\right)^2+(1-\theta)\right]^k\\ &= A_l^{(0)}\left[\theta\left(\tanh(\beta_l b)\right)^2+(1-\theta)\right]^{k+1}, \end{aligned} \tag{31}$$

which proves (23) for $n=k+1$. From problem (8), we have for $n=k+1$

$$w^{(2(k+1)+1)}(x,y)=\sum_{m=1}^{\infty}A_m^{(2(k+1)+1)}\sin\left(\beta_m x\right)\sinh\left(\beta_m y\right) \tag{32}$$

Using the boundary condition $\partial_\nu w|_{\Gamma_0}^{(2(k+1)+1)}=\partial_\nu w|_{\Gamma_0}^{(2(k+1))}$ we obtain

$$\sum_{m=1}^{\infty}A_m^{(2(k+1)+1)}\sin\left(\beta_m x\right)\beta_m\cosh\left(\beta_m b\right)=\sum_{m=1}^{\infty}A_m^{(2(k+1))}\sin\left(\beta_m x\right)\beta_m\sinh\left(\beta_m b\right)$$

Again, Multiplication by $\sin(\beta_l x)$, integration over the interval $(0,a)$ and the orthogonality (27) lead to

$$A_l^{(2(k+1)+1)} = A_l^{(2(k+1))} \tanh\left(\beta_n b\right), \quad \text{for } l = 1, 2, \ldots$$

The introduction of the expression of $A_l^{(2(k+1))}$ defined in (31) in this last equality gives the formula (24) for $n = k + 1$ and this completes the proof of the lemma. □

Now, we return to the proof of the theorem.

Proof of (1) of the Theorem 1 Recall that from the Lemma 1, we just need to show the convergence of $w^{(2n)}|_{\Gamma_0}$ to zero. So let us consider the norm of $w^{(2n)}$ in $\mathcal{L}^2(\Gamma_0)$:

$$\| w^{(2n)}(x, b) \|_{\mathcal{L}^2(\Gamma_0)} = \left(\int_0^a |w^{(2n)}(x, b)|^2 dx \right)^{1/2},$$

By Parsaval's identity we have

$$\int_0^a |w^{(2n)}(x, b)|^2 dx = \frac{a}{2} \sum_{m=1}^{\infty} \left[A_m^{(0)} \left[\theta\left(\tanh(\beta_m b)\right)^2 + (1-\theta) \right]^n \cosh(\beta_m b) \right]^2,$$

thus

$$\| w^{(2n)}(x, b) \|^2_{\mathcal{L}^2(\Gamma_0)} - \frac{a}{2} \sum_{m=1}^{\infty} (A_m^{(0)})^2 \left[\theta\left(\tanh(\beta_m b)\right)^2 + (1-\theta) \right]^{2n} \left(\cosh(\beta_m b)\right)^2,$$

and hence

$$\| w^{(2n)}(x, b) \|^2_{\mathcal{L}^2(\Gamma_0)} - \sum_{m=1}^{\infty} \hat{A}_m^{(0)} \left[\theta\left(\tanh(\beta_m b)\right)^2 + (1-\theta) \right]^{2n}$$

where $\hat{A}_m^{(0)} = \left(\frac{a}{2}\right)\left(A_m^{(0)}\right)^2 \left(\cosh(\beta_m b)\right)^2$. Therfore we obtain that $\lim_{n\to\infty} \| w^{(2n)} \| = 0$ when

$$| \theta\left(\tanh(\beta_m b)\right)^2 + (1-\theta) | < 1, \ \forall m \geq 1$$

i.e.

$$-1 < \theta\left(\tanh(\beta_m b)\right)^2 + (1-\theta) < 1, \ \forall m \geq 1$$

guiving

$$0 < \theta < \frac{2}{1 - \left(\tanh(\beta_m b)\right)^2}, \ \forall m \geq 1. \tag{33}$$

Frome $\beta_m = \frac{m\pi}{a}$ we get that $\tanh(\beta_1 b) \leq \tanh(\beta_m b), \ \forall m \geq 1$. Thus (33) is equivalent to

$$0 < \theta < \frac{2}{1 - \left(\tanh(\beta_1 b)\right)^2}$$

i.e. the convergence of $w^{(2n)}|_{\Gamma_0}$ to zero is ensured when $\theta \in (0, \theta^*)$ with $\theta^* = \dfrac{2}{1-\left(\tanh(\beta_1 b)\right)^2}$ so the proof of the theorem is achieved. □

4 Acceleration of Convergence

Note that even if the JN relaxed algorithm is widely used in works dealing with applications modeled by Cauchy problems, there is no estimate of the interval in which the relaxation parameter must be chosen to accelerate convergence. We propose in the following theorem an identification of this convergence acceleration interval.

Theorem 2 *Considering the Poisson problem defined in (2)–(4). We have: The convergence of the JN relaxation algorithm is more faster than the convergence of the KMF algorithm* $\forall\theta \in]1, \theta_a[$ *with* $\theta_a = \frac{1+\left(\tanh(\frac{\pi}{a}b)\right)^2}{1-\left(\tanh(\frac{\pi}{a}b)\right)^2}$

Proof Recall that the standard KMF algorithm [6] can be obtained from the JN relaxated algorithm [11] by taking the relaxation parameter $\theta = 1$.

Denote by $\{w_\theta^{(n)}\}$ the sequence solutions obtained from the JN relaxation algorithm for the parameter θ.

Note that

$$w_\theta^{(2n)}(x, y) = \sum_{m=1}^{\infty} A_m^{(0)} \left[\theta\left(\tanh(\beta_m b)\right)^2 + (1-\theta)\right]^n \sin\left(\beta_m x\right)\cosh\left(\beta_m y\right) \quad (34)$$

Thus, for $\theta = 1$, we obtain

$$w_1^{(2n)}(x, y) = \sum_{m=1}^{\infty} A_m^{(0)} \left[\left(\tanh(\beta_m b)\right)^2\right]^n \sin\left(\beta_m x\right)\cosh\left(\beta_m y\right) \quad (35)$$

To find the condition for which the convergence of $\{w_\theta^{(n)}\}$ is faster than the convergence of the sequence $\{w_1^{(n)}\}$, we must verify the following condition

$$|\, \theta\left(\tanh(\beta_m b)\right)^2 + (1-\theta)\, | < \left(\tanh(\beta_m b)\right)^2, \ \forall m \geq 1$$

so

$$-\left(\tanh(\beta_m b)\right)^2 - 1 < \theta\Big(\left(\tanh(\beta_m b)\right)^2 - 1\Big) < \left(\tanh(\beta_m b)\right)^2 - 1, \ \forall m \geq 1.$$

Dividing by $\Big(1 - \left(\tanh(\beta_m b)\right)^2\Big)$, which is positive, we obtain

$$-\frac{\left(1+\left(\tanh(\beta_m b)\right)^2\right)}{1-\left(\tanh(\beta_m b)\right)^2} < -\theta < -1, \ \forall m \geq 1.$$

Which gives

$$\frac{1+\left(\tanh(\beta_m b)\right)^2}{1-\left(\tanh(\beta_m b)\right)^2} > \theta > 1, \ \forall m \geq 1.$$

Since $\tanh(\beta_1 b) \leq \tanh(\beta_m b), \ \forall m \geq 1$, these inequalities are equivalent to

$$1 < \theta < \frac{1+\left(\tanh(\beta_1 b)\right)^2}{1-\left(\tanh(\beta_1 b)\right)^2}.$$

This completes the proof of the Theorem 2. □

5 Numerical Results

In this section we present some numerical experiments to show the efficiency of the proposed approach. We consider a bounded domain in $\mathbb{R}^2$ with tow boundaries, one is accessible, while the other is not. This means that we suppose that $\Gamma_2 = \Gamma_3 = \emptyset$.

Let's $\Omega = \left\{(x, y) \in \mathbb{R}^2, \ 0.5 < x^2 + y^2 < 1\right\}$, with boundary $\partial\Omega = \Gamma_0 \cup \Gamma_1$. Such that $\Gamma_0 = \left\{x^2 + y^2 = 0.25\right\}$ is the inaccessible boundary while $\Gamma_1 = \left\{x^2 + y^2 = 0.25\right\}$ is the accessible one where the measures are done.

We consider the following data

$$f = -(a_1^2 + a_2^2)\exp(a_1 x + a_2 y), \ \ g{=}(a_1 * n_x + a_2 * n_y)\exp(a_1 x + a_2 y), \ \ f_1 = \exp(a_1 x + a_2 y),$$

where n_x and n_y are the X-axis and the Y-axis component respectively, and $a_1 = 0.5$, $a_2 = \sqrt{(1 - a_1^2)}$.

The exact solution of Cauchy Poisson problem corresponds to these data is defined by

$$ue(x, y) = \exp(a_1 x + a_2 y).$$

We start by presenting the evolution of the number of the iteration according to the relaxation parameter θ. In this case we fix the stopping criterion and the vary the parameter θ, then we keep the number of the iteration at the convergence. Thus, in the Fig. 1 we can observe the confirmation of the the theoretical result which clams that there exists a θ^* which depends on the domain such that $\forall\, \theta \in (0, \theta^*)$ we have the convergence. Moreover, we can clearly see the acceleration of the procedure according to θ. We choose $\theta = 1.8$ and we run the algorithm. The comparison between the approximate and the exact data on the inaccessible boundary Γ_0 is

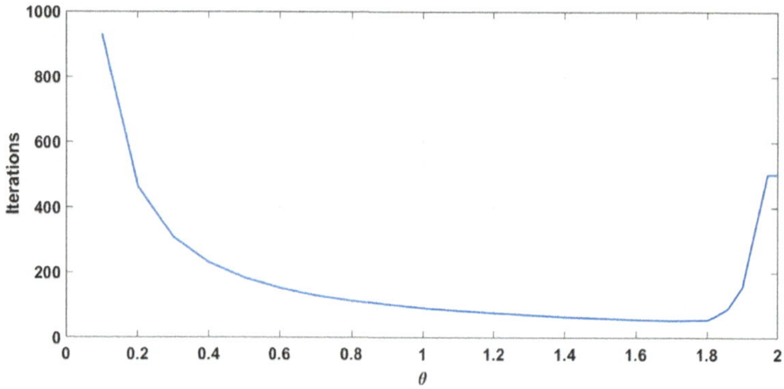

Fig. 1 Evolution of the number of iteration at the convergence according to θ

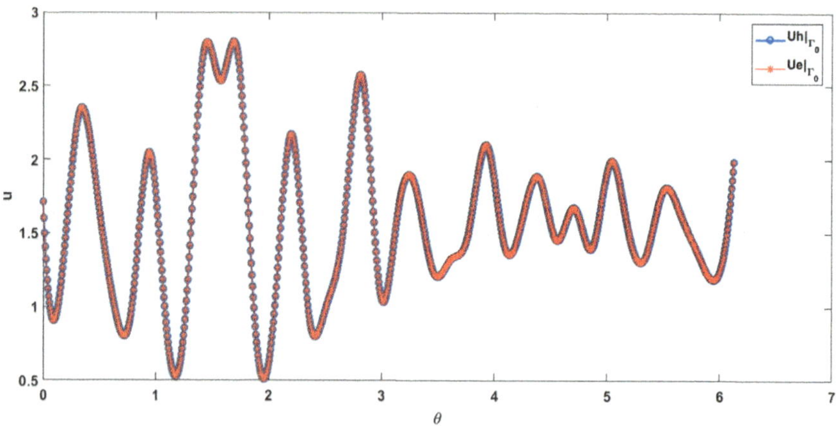

Fig. 2 Comparison between the approximate and the exact data on Γ_0

presented in the Fig. 2. As we see the two curves correspond to approximate and exact data respectively are superimposed.

In order to discuss the stability of the proposed algorithm we present some obtained results with different level of the added noise to the given data. Let's δ be the level of the noise. The perturbed data is given by

$$g = g + \delta(2r - 1)g, \quad f_1 = f_1 + \delta(2r - 1)f_1$$

where r is a Gaussian noise normally distributed on [0, 1]. In the Fig. 3 we present the comparison between the approximate and the exact data on Γ_0 according to the different noise level $\delta \in \{5\%, 20\%, 30\%\}$. The figure confirms the stability of the algorithm even when the noise level is very high. Finally in the Fig. 4 we present the difference between the obtained solution and the exact one for each noise level.

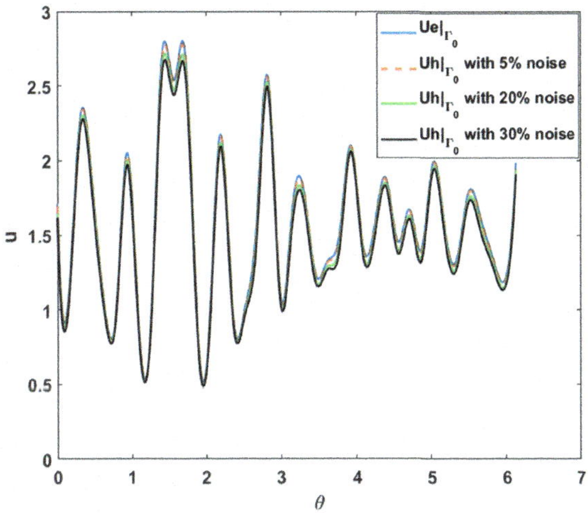

Fig. 3 Comparison between the approximate and the exact data on Γ_0 according to the different noise level $\delta \in \{5\%, 20\%, 30\%\}$

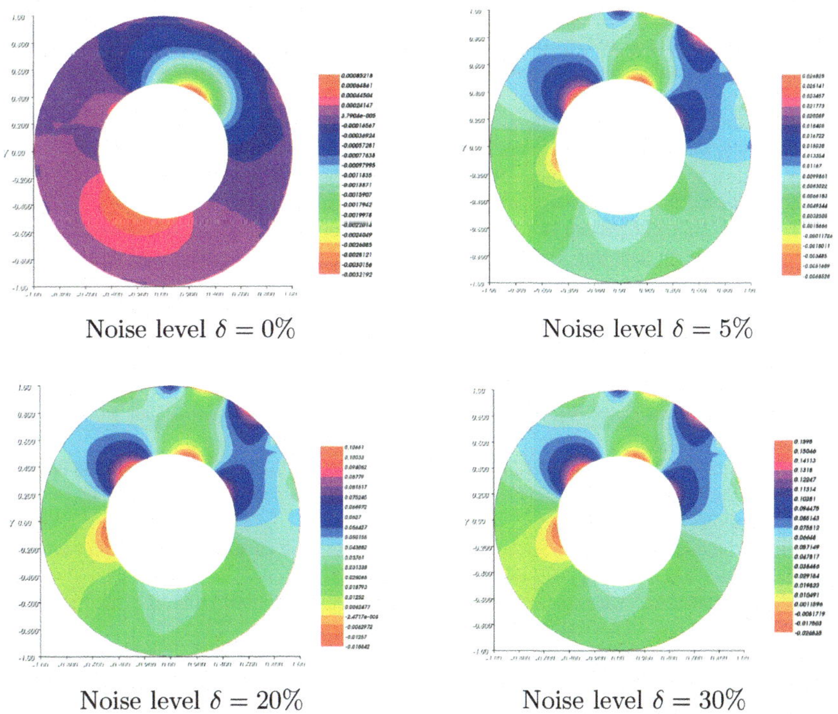

Fig. 4 Difference between the obtained solution and the exact one for each noise level $\delta \in \{5\%, 20\%, 30\%\}$

6 Conclusion

In this work a new convergence result of the relaxed iterative JN algorithm for solving the inverse elliptic Cauchy problem has proposed. Unlike to the preview works an effective interval which only depends on the problem data is given. Through some presented numerical experiments we have confirmed the theoretical results. The proposed approach can easily extended to the other elliptic inverse Cauchy problems.

References

1. A. Bergam, A. Chakib, A. Nachaoui, M. Nachaoui, Adaptive mesh techniques based on a posteriori error estimates for an inverse cauchy problem, Applied Mathematics and Computation 346 (2019) 865–878
2. F. Berntsson, V. A. Kozlov, L. Mpinganzima, B.-O. Turesson, Iterative tikhonov regularization for the cauchy problem for the helmholtz equation, Computers & Mathematics with Applications 73 (1) (2017) 163–172
3. A. Chakib, A. Nachaoui, M. Nachaoui, H. Ouaissa, On a fixed point study of an inverse problem governed by stokes equation. Inverse Problems **35**(1), 015008 (2018)
4. L. Caillé, L. Marin, F. Delvare, A meshless fading regularization algorithm for solving the cauchy problem for the three-dimensional helmholtz equation, Numerical Algorithms 82 (3) (2019) 869–894
5. A. Ellabib, A. Nachaoui, An iterative approach to the solution of an inverse problem in linear elasticity, Mathematics and Computers in Simulation 77 (2–3) (2008) 189–201
6. V. Kozlov, V. Maz'ya, A. Fomin, An iterative method for solving the cauchy problem for elliptic equm'ions£, 9
7. L. Marin, B. T. Johansson, A relaxation method of an alternating iterative algorithm for the cauchy problem in linear isotropic elasticity, Computer methods in applied mechanics and engineering 199 (49–52) (2010) 3179–3196
8. A. Nachaoui, M. Nachaoui, T. Tadumadze, Electrical potentials measured on the surface of the knee for detecting osteoarthritis-induced cartilage degeneration, in: Second International Conference of Mathematics in Erbil (SICME2019) (2019)
9. A. Nachaoui, M. Nachaoui, Iterative methods for forward and inverse bioelelectric field problem, in: International Conference on Applied Mathematics, Modeling and Life Sciences, Icamls' 18 (2018)
10. M. Jourhmane, A. Nachaoui, A relaxation algorithm for solving a cauchy problem, in: Proceedings of the Second International Conferences on Inverse Problems in Engineering, Engineering Foundation (1996)
11. M. Jourhmane, A. Nachaoui, An alternating method for an inverse cauchy problem, Numerical Algorithms 21 (1) (1999) 247–260
12. M. Jourhmane, A. Nachaoui, Convergence of an alternating method to solve the cauchy problem for poisson's equation, Applicable analysis 81 (5) (2002) 1065–1083
13. F. Aboud, A. Nachaoui, M. Nachaoui, On the approximation of a cauchy problem in a nonhomogeneous medium, in: Journal of Physics: Conference Series, Vol. 1743, IOP Publishing, p. 012003 (2021)
14. K.A. Berdawood, A. Nachaoui, M. Nachaoui, F. Aboud, An effective relaxed alternating procedure for cauchy problem connected with helmholtz equation. Numeri. Methods Partial Differ. Equ.

15. K. Berdawood, A. Nachaoui, R. Saeed, M. Nachaoui, F. Aboud, An efficient dn alternating algorithm for solving an inverse problem for helmholtz equation. Discrete Contin. Dyn. Syst.
16. K. A. Berdawood, A. Nachaoui, R. Saeed, M. Nachaoui, F. Aboud, An alternating procedure with dynamic relaxation for cauchy problems governed by the modified helmholtz equation, Advanced Mathematical Models & Applications 5 (1) (2020) 131–139

Identification of Genuine from Fake Banknotes Using an Enhanced Machine Learning Approach

S. Lyaqini and M. Nachaoui

Abstract In this study, we propose a supervised learning problem to differentiate genuine from fake banknotes. The problem is reformulated as a regularized optimization problem whose fidelity term is the hinge loss function and the hypothesis space is constructed by a reproducing kernel Hilbert space (RKSH). The numerical results are realized by the Adam method, which shows her success in detecting counterfeit banknotes compared to the stochastic gradient descent (SGD) algorithm.

Keywords Banknote authentication · Smooth approximation · Hing loss · Supervised problem · Tikhonov regularization

1 Introduction

Evolving technologies are making great contributions to human life in various fields under the name of artificial intelligence. One of these domains is the financial security, more precisely the detection of counterfeit banknotes. It is difficult for humans to differentiate between real and counterfeit bills, especially since they have many similar characteristics. In studies from recent years, prediction systems based on machine learning have used Artificial Neural Network, Support Vector Machine and K Nearest Neighbours classifiers for banknote recognizing [1–3].

More concretely, our contribution this paper is the consideration of machine learning approaches to evaluate authentication of banknotes. More precisely, the problem is reformulated as a regularized optimization problem whose fidelity term is the hinge loss function and the hypothesis space is constructed by a reproducing kernel Hilbert space (RKSH) [4]. Although, it is well known that reproducing kernel Hilbert space have become increasingly important in machine learning tasks and supervised learning problem with kernel has attracted more and more attentions in machine learning. The most important tools in RKSH is the representer theorem [5], which states that any function in an RKHS that minimizes an empirical risk function can be written as a linear combination of the kernel function evaluated at the

S. Lyaqini (✉) · M. Nachaoui
LMA FST Beni-Mellal, University Sultan Moulay Slimane, Slimane, Morocco

A. Nachaoui et al. (eds.), *Mathematical Control and Numerical Applications*,
Springer Proceedings in Mathematics & Statistics 372,
https://doi.org/10.1007/978-3-030-83442-5_5

training set. This is a useful result in practice, as it effectively simplifies the problem of empirical risk minimization from an infinite-dimensional to a finite dimensional optimization problem. In addition, the choice of the loss function is explained by the fact that it penalizes predictions not only when they are incorrect, but also when they are correct but not trustful. It penalizes severely wrong predictions significantly, correct but insecure predictions a little less, and only correct and secure predictions are not penalized at all. However, the fidelity term of the resulting minimization problem is not differentiable, using the smooth approximation techniques [6–8] we transform it into a smoothed one, which is solved directly using Adam algorithm [9, 10]. Lastly, we present several numerical validations of the proposed approach and a comparison with stochastic gradient descent (SGD) algorithm [11]. In specific, the numerical results realized on banknote authentication data set indicate that the proposed approach is an efficient and helpful tool to differentiate genuine from fake banknotes.

We briefly summarize the contents of the paper. In Sect. 2, we present the setting of the classification problem and its reformulation as a minimization problem, using hinge loss function. Section 3 is concerned to a smooth approximation of the hinge loss function. In Sect. 4, we evaluate the efficiency of the proposed approach on banknote authentication data set compared to stochastic gradient descent (SGD) algorithm to differentiate genuine from fake banknotes.

2 Formulating the Binary Classification Problem

Given a training data set $\mathcal{D} = \{(x_i, y_i)\}_{i=1}^n$, where $x_i \in \mathcal{X} \subset \mathbb{R}^d$ called the input space and $y_i \in \mathcal{Y} = \{-1, 1\}$ is the output space. The aim of binary classification is to classify observations that take values in $\mathcal{X}$ into one of two classes, -1 or $+1$. A discriminant model $f : \mathcal{X} \longrightarrow \mathcal{Y}$ yields a classifier $sgn(f(x)) \in \mathcal{Y}$ that represents our guess of the label $\mathcal{Y}$ of a future observation $\mathcal{X}$. Therefore, the optimal decision can be obtained by solving the following regularized minimization problem.

$$\min_{f \in \mathcal{K}} \frac{1}{n} \sum_{i=1}^{n} V(f(x), y) + C\|f\|_{\mathcal{S}}^2, \tag{1}$$

where $\mathcal{K}$ denotes a reproducing kernel Hilbert space [4], $\|.\|_{\mathcal{S}}$ the corresponding norm, C is a positive regularization parameter and V is a loss function penalizing the deviation between $f(x_i)$ and y_i. Throughout this paper we use the hinge loss given by

$$V(\omega, v) = \max(1 - \omega v, 0).$$

2.1 Kernel Function

A kernel $\mathcal{S} : \mathcal{X} \times \mathcal{X} \longrightarrow \mathbb{R}$ is a function, such that for all $x, y \in \mathcal{X}$,

$$\mathcal{S}(x, y) = \langle \Phi(x), \Phi(y) \rangle$$

where Φ is a mapping from $\mathcal{X}$ to a (inner product) feature space (high dimensional space) $\mathcal{K}$ and $\langle \Phi(x), \Phi(y) \rangle$ denotes the inner product of $\Phi(x)$ and $\Phi(y)$.

A kernel function must satisfy the following two properties:

(i) Symmetricity: $\mathcal{S}(x, y) = \mathcal{S}(y, x)$
(ii) Cauchy-Schwarz inequality: $\left(\mathcal{S}(x, y)\right)^2 \leq \mathcal{S}(x, x)\mathcal{S}(y, y)$

Let $\mathcal{K}$ a Hilbert space of real valued functions defined on $\mathcal{X}$ and equipped with an inner product $\langle \cdot, \cdot \rangle$. Then the kernel $\mathcal{S}(x, y)$ is said to be positive definite if, for any finite point set $x_1, ..., x_n \in \mathcal{X}$ and for any $a_1, ..., a_n \in \mathbb{R}$

$$\sum_{i,j=1}^{n} a_i a_j \mathcal{S}(x_i, x_j) \geq 0,$$

Any positive definite kernel $\mathcal{S}(x, y)$ is a reproducing kernel of $\mathcal{K}$ if it verifies the following properties:

(i) $\mathcal{S}_x(\cdot) = \mathcal{S}(\cdot, x) \in \mathcal{K}, \quad \forall x \in \mathcal{X}$
(ii) $f(x) =< f, \mathcal{S}_x >, \quad \forall f \in \mathcal{K}$

Such a Hilbert space $\mathcal{K}$ is called a reproducing kernel Hilbert space (RKHS).

3 Hinge Loss Function

The hinge loss functions is non differentiable and cannot be used with optimization methods which required the assumption of differentiability of the objective function. To overcome this problem the hinge loss function can be accurately approximated by a smooth function which is twice differentiable and convex [8, 12, 13] defined below. For all $u \in \mathbb{R}$, the max function is approximated by a smooth function,

$$\max(u, 0) = \left(u + \frac{1}{\gamma} \log\left[1 + e^{(-\gamma u)}\right]\right), \qquad \forall \gamma > 0.$$

We first notice that the loss function is always a true function of only one variable u, with $u = 1 - \omega v$.

Let us denote by $V_\gamma(u)$ the smoothed loss function with parameter γ of $V(u) = \max(u, 0)$ defined by

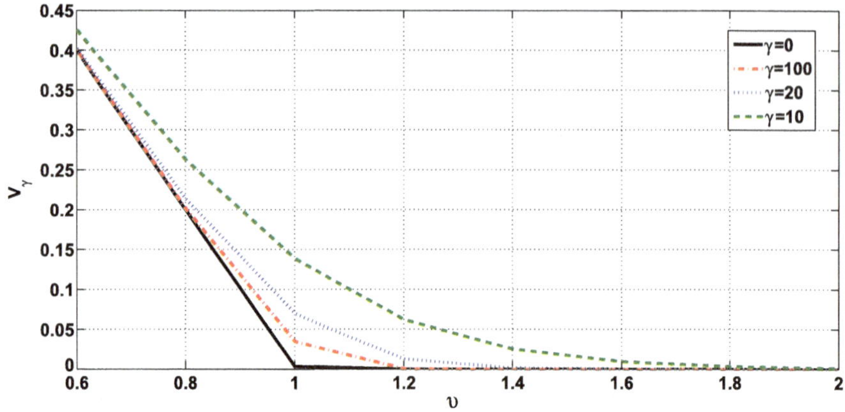

Fig. 1 Smoothed hings loss

$$V_\gamma(u) = \left(u + \frac{1}{\gamma}\log\left[1 + e^{(-\gamma u)}\right]\right), \qquad \forall\gamma > 0.$$

The Lemma 2 bounds the difference between hinge loss function, V, and its smooth approximation V_γ (Fig. 1).

Lemma 1 *For all* $u \in \mathbb{R}$, *then*

$$|V_\gamma(u) - V(u)| \leq \frac{\log(2)}{\gamma} \tag{2}$$

Proof We consider three cases.
For $u > 0$, we have that $V(u) = u$, then

$$\begin{aligned} |V_\gamma(u) - V(u)| &= |u + \frac{1}{\gamma}\log\left[1 + e^{(-\gamma u)}\right] - u| \\ &= \frac{1}{\gamma}\log\left[1 + e^{(-\gamma u)}\right] \\ &\leq \frac{\log(2)}{\mu}. \end{aligned}$$

Since $u \mapsto \frac{1}{\gamma}\log\left[1 + e^{(-\gamma u)}\right]$ is an decreasing function and $u \geq 0$.
For $u \leq 0$, we have that $V(u) = 0$, then

$$\begin{aligned} |V_\gamma(u) - V(u)| &= |V_\gamma(u)| \\ &\leq V_\gamma(0) = \frac{\log(2)}{\mu}. \end{aligned}$$

Since V_γ is an increasing function and $u \leq 0$, then $V_\gamma(u) \leq V_\gamma(0)$.

By Lemma 2, we have that as γ tends to $+\infty$, the smoothed hinge loss V_γ approaches the hinge loss V. The V_γ is used here to replace the fidelity term $\mathcal{I}[f] := \frac{1}{n}\sum_{i=1}^{n} V(1 - f(x_i)y_i)$ of the problem (1) by

$$\mathcal{I}^\gamma[f] := \frac{1}{n}\sum_{i=1}^{n} V_\gamma(1 - f(x_i)y_i),$$

which is twice differentiable and convex. Using Lemma 2 we shall show that the $\mathcal{I}^\gamma$ converges to the empirical risk $\mathcal{I}$ as the smoothing parameter γ go to infinity.

Lemma 2 *For all $f \in \mathcal{K}$, then*

$$|\mathcal{I}^\gamma[f] - \mathcal{I}[f]| \leq \frac{\log(2)}{\mu}. \tag{3}$$

Proof For all $f \in \mathcal{K}$, we have

$$V_\gamma(1 - f(x_i)y_i) - V(1 - f(x_i)y_i) \leq \frac{\log(2)}{\mu}, \quad \forall i = 1, ..., n.$$

Then

$$\frac{1}{n}\sum_{i=1}^{n} V_\gamma(1 - f(x_i)y_i) - \frac{1}{n}\sum_{i=1}^{n} V(1 - f(x_i)y_i) \leq \frac{1}{n}\sum_{i=1}^{n} \frac{2\log(2)}{\mu}.$$

Thus

$$0 \leq \mathcal{I}^\gamma[f] - \mathcal{I}[f] \leq \frac{\log(2)}{\mu}.$$

Since $\mathcal{I}^\gamma[f] \geq \mathcal{I}[f]$.
Therefore

$$|\mathcal{I}^\gamma[f] - \mathcal{I}[f]| \leq \frac{\log(2)}{\mu}.$$

In the sequel, we consider the approximate optimization problem of (1) given by

$$\min_{f \in \mathcal{K}} \frac{1}{n}\sum_{i=1}^{n} V_\gamma(f(x), y) + C\|f\|_{\mathcal{S}}^2. \tag{4}$$

Thus, the objective function in (4) is differentiable.

According to the representer theorem [5], The solution f^γ of the problem (4), can be written as a finite linear combination of kernel evaluations in the data, namely

$$f^{\gamma}(x) = \sum_{i=1}^{n} \alpha_i \mathcal{S}(x, x_i), \tag{5}$$

where $\alpha_i \in \mathbb{R}$, $i = 1, ..., n$ and $\mathcal{S}$ the reproducing kernel of $\mathcal{K}$. Therefore, the solution to the possibly infinite dimensional optimization problem (4) resides in the n-dimensional span of the functions $\mathcal{S}_{x_i}(.)$, $i = 1, ..., n$. To compute the coefficients, we simply have to solve the problem

$$\min_{\alpha \in \mathbb{R}^n} \varphi_C^{\gamma}(\alpha), \tag{6}$$

where

$$\varphi_C^{\gamma}(\alpha) = \frac{1}{n}\sum_{i=1}^{n}\left(1 - \alpha^\top \mathcal{S}_i y_i + \frac{1}{\gamma}\log\left(1 + e^{(-\gamma(1-\alpha^\top \mathcal{S}_i y_i))}\right)\right) + C\alpha^\top \mathcal{S}\alpha,$$

Denoting $\mathcal{S} = \begin{pmatrix} \mathcal{S}(x_1, x_1) & \cdots & \mathcal{S}(x_1, x_n) \\ \vdots & \ddots & \vdots \\ \mathcal{S}(x_n, x_1) & \cdots & \mathcal{S}(x_n, x_n) \end{pmatrix}$, $\mathcal{S}_i = \begin{pmatrix} \mathcal{S}_{x_i}(x_1) \\ \vdots \\ \mathcal{S}_{x_i}(x_n) \end{pmatrix}$, $\alpha = \begin{pmatrix} \alpha_1 \\ \vdots \\ \alpha_n \end{pmatrix}$ and $y = \begin{pmatrix} y_1 \\ \vdots \\ y_n \end{pmatrix}$.

Thanks to the differentiability of the objective function of (6), we use a Adam algorithm to solve this reformulation [9, 10].

In the next section, we will treat the experimental results obtained after the simulation of banknote authentication data set using the proposed approach.

4 Experimental Results

To evaluate this algorithm it is often desirable to have some standardized benchmark data sets. In our case we choose to evaluate the proposed method through real-life data from the UCI machine learning repository. To evaluate the efficiency of the proposed algorithm, we compared its accuracy with the Stochastic gradient descent algorithm. The goal is to show that the proposed approach is faithfully and faster to predict the considered class and we investigate on the importance of the features to find out which features contribute to most towards achieving the result.

Throughout this paper we use the Gaussian kernel, given by

$$\mathcal{S}(x, x') = e^{\left(-\frac{1}{2\sigma^2}\sum_{i=1}^{n}(x_i - x_i')^2\right)},$$

Table 1 Banknote authentication data set description

Attributes	Description	Type of data
Variance	Variance of Wavelet Transformed Image	Digital
Skewness	Skewness is the measure of the lack of symmetry	Digital
Kurtosis	Kurtosis is a measure of whether the data are heavy-tailed or light-tailed relative to a normal distribution	Digital
entropy	Image entropy is a quantity which is used to describe the amount of information which must be coded for, by a compression algorithm	Digital
Class	Class contains two values 0 representing genuine note and 1 representing fake note	Digital

as reproducing kernel of the space $\mathcal{K}$, where σ is a free parameter and n is the size of training set.

The data set used to train the models is taken from UCI machine learning repository [14]. Data were extracted from genuine and counterfeit banknote images. The dataset has 1372 instances. There are 5 attributes out of which 4 are the features and one is the target attribute. The dataset contains a balanced ratio of both classes which is $55 : 45$ (genuine: counterfeit). The target class contains two values: 0 and 1 where 0 represents genuine note and 1 represents fake note.

The first 4 columns have been normalized to have mean 0 and Euclidean norm 1. For the implementation, we construct our training set by taking 1097 random observations form all data sets. Then by using the proposed approach, we generate predicting model. Let's first find the correlation of each of these feature pairs and visualize the correlations using a heatmap (Figs. 2 and 3).

As it can be seen, from figures below that Adam algorithm trained the model with high accuracy unlike the SGD algorithm.

In Table 1, we present the accuracy to measure the performance of the models implemented (Figs. 4, 5 and 6).

The accuracy of the test is its ability to differentiate the genuine and fake note test cases correctly.

$$\frac{TP + TN}{TP + TN + FP + FN}$$

The sensitivity of a test is its ability to determine the genuine note cases correctly.

$$\frac{TP}{TP + FN}$$

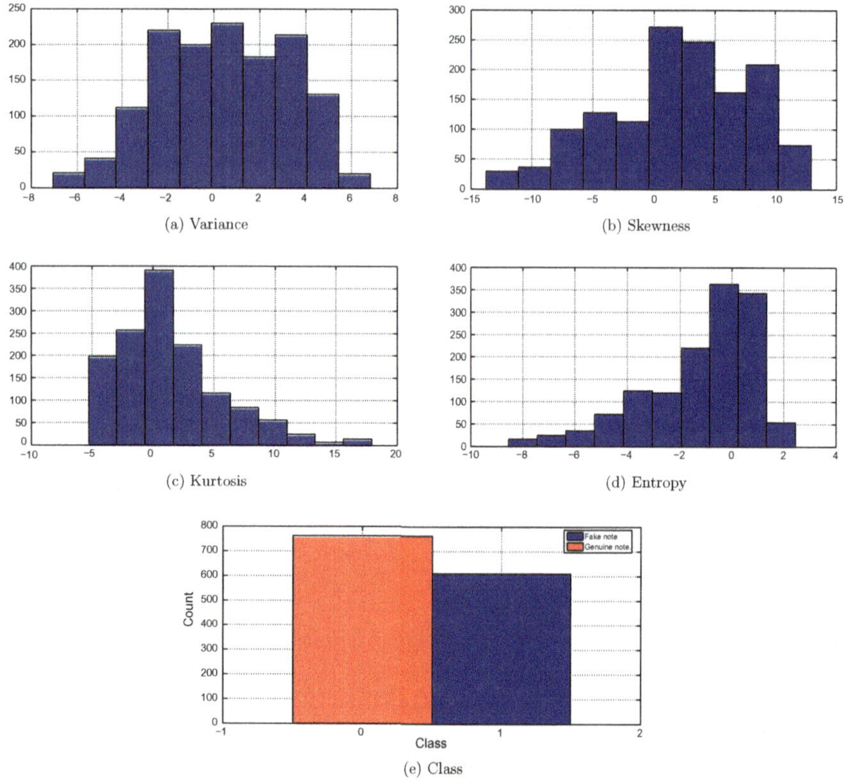

Fig. 2 Banknote authentication data set information

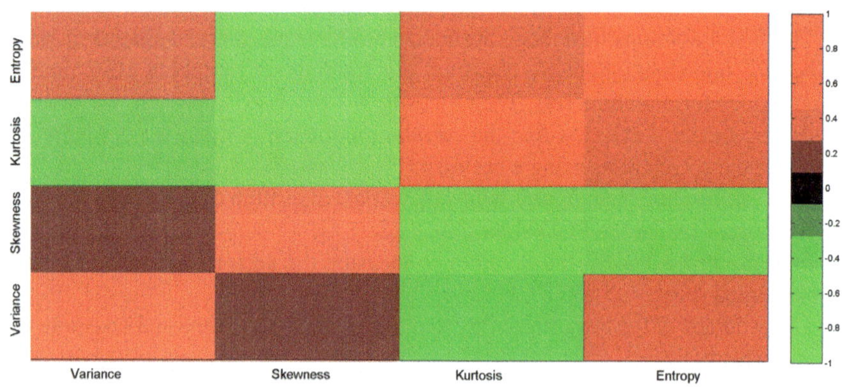

Fig. 3 Correlation between the features in the banknote dataset

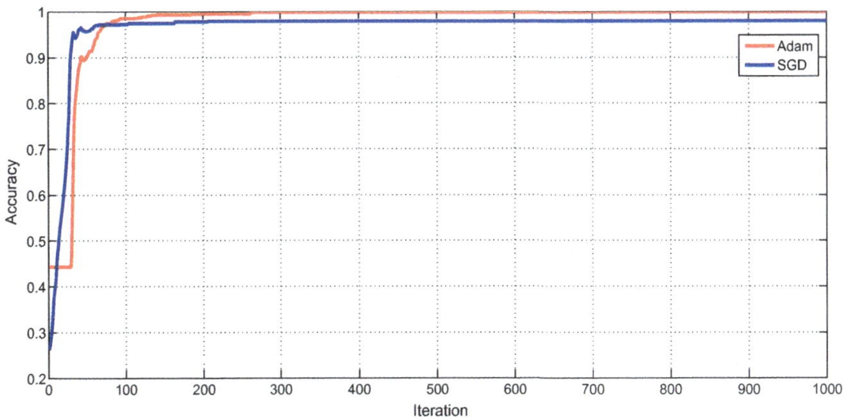

Fig. 4 Accuracy

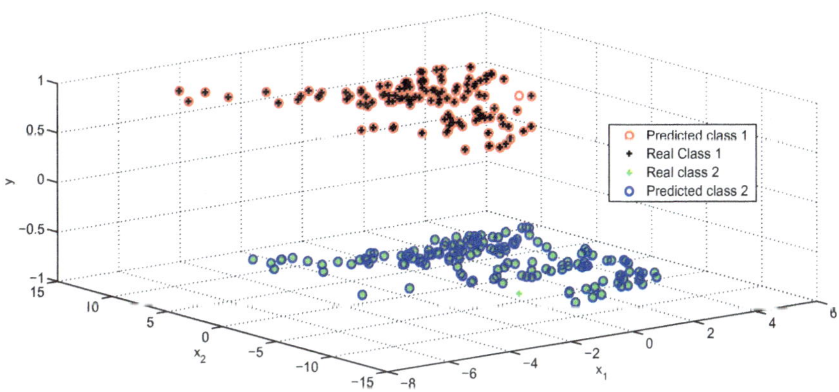

Fig. 5 Real and predicted model using Adam algorithm, for $\sigma = 3$, $\alpha = 10$ and $\lambda = 10^{-5}$

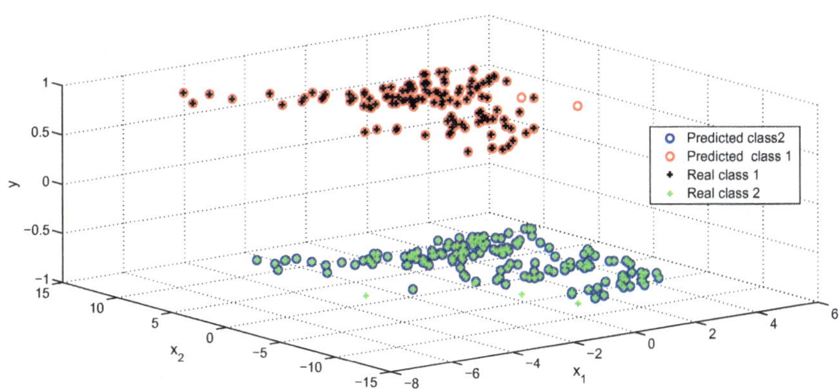

Fig. 6 Real and predicted model using Stochastic gradient descent algorithm, for $\sigma = 3$, $\alpha = 10$ and $\lambda = 10^{-5}$

Table 2 The accuracy obtained by Adam and SGD methods

	Adam (%)	SGD (%)
Accuracy	99.63	97.81
Sensitivity	100	100
Specificity	99.13	95.79
Precision	99.37	96.89

The specificity of a test is its ability to determine the fake note cases correctly.

$$\frac{TN}{TN + FP}$$

The precision of a test is its ability to determine the number of notes that classifier labeled as genuine is actually genuine

$$\frac{TP}{TP + FP}$$

where,

- True Positive (TP) = the number of cases correctly identified as genuine notes.
- True negative (TN) = the number of cases correctly identified as fake notes.
- False positive (FP) = the number of cases incorrectly identified as genuine notes.
- False negative (FN) = the number of cases incorrectly identified as fake notes.

The results presented in this table show that the proposed approach is not only the best on the training set but also has a very good testing set accuracy. As a conclusion, we can say that the proposed methods is became more efficient in detecting fake banknotes (Table 2).

From Fig. 7a, it is clear that for the number of correctly classified instances is 274, and the number of misclassified instances is 1. From Fig. 7b, it is clear that for the number of correctly classified instances is 269, and the number of misclassified instances is 5. As there are just 1 misclassified patterns using Adam algorithm, the model is predicting this testing data very well unlike SGD algorithm.

4.1 Feature Importance

In this section, we investigate on the importance of the features to find out which features contribute to most towards achieving the result. Furthermore, we select the two features which are the most important to build our predictive model.

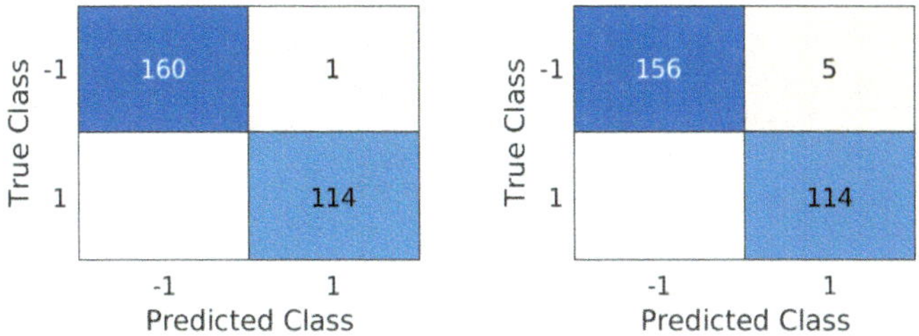

(a) Adam (b) SGD

Fig. 7 Confusion matrix

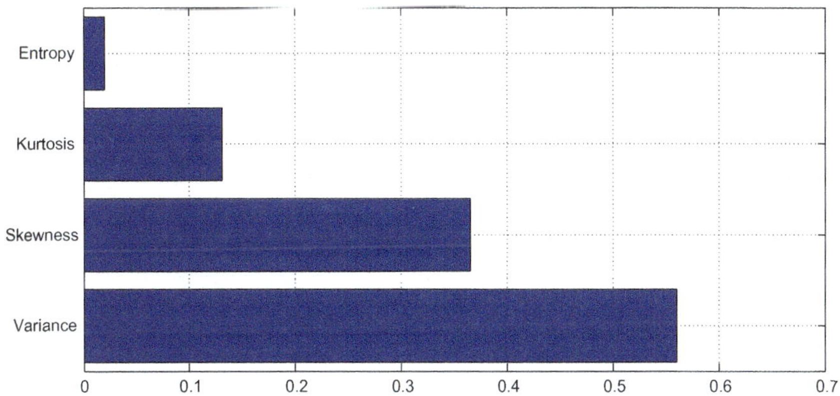

Fig. 8 Feature importance

From Fig. 8, we can clearly see that, feature 1 (**variance**) is the most important feature. Followed by feature 2 (**Skewness**). In the following, we also trained and tested by using only the features, variance and skewness.

Figure 9, shows that the obtained model is also better at detection of counterfeit banknotes.

5 Conclusion

In this work we have constructed a model to detect the banknote as genuine or fake using the supervised learning problem based on smoothed hing loss. Extensive experiments have been executed on banknote authentication using Adam algorithm to find the best suitable model for banknote recognition. The experimental results shows that our proposed approach outperforms the stochastic gradient descent and gives a good success rate.

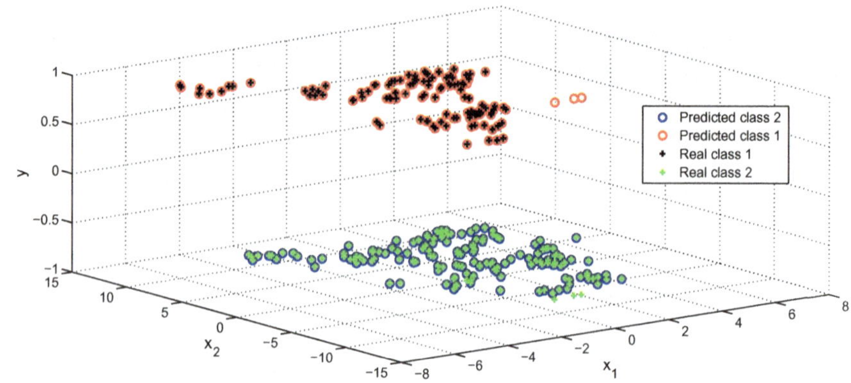

Fig. 9 Sparsity Example: Fitting only features **variance** and **Skewness**

References

1. M. Leo, G. Medioni, M. Trivedi, T. Kanade, G. M. Farinella, Computer vision for assistive technologies, Computer Vision and Image Understanding 154 (2017) 1–15
2. J. W. Lee, H. G. Hong, K. W. Kim, K. R. Park, A survey on banknote recognition methods by various sensors, Sensors 17 (2) (2017) 313
3. A. Ghazvini, J. Awwalu, A. A. Bakar, Comparative analysis of algorithms in supervised classification: A case study of bank notes dataset, Int. J. Comput. Trends Technol 17 (1) (2014) 39–43
4. N. Aronszajn, Theory of reproducing kernels, Transactions of the American Mathematical Society 68 (1950) 337–404
5. B. Schölkopf, R. Herbrich, A. J. Smola, A generalized representer theorem, in: International Conference on Computational Learning Theory, Springer, pp. 416–426 (2001)
6. Y.-J. Lee, W.-F. Hsieh, C.-M. Huang, epsilon-ssvr: a smooth support vector machine for epsilon-insensitive regression. IEEE Trans. Knowl. Data Eng. (5), 678–685 (2005)
7. J. Hajewski, S. Oliveira, D. Stewart, Smoothed hinge loss and ℓ^1 support vector machines, in: 2018 IEEE International Conference on Data Mining Workshops (ICDMW), IEEE, pp. 1217–1223 (2018)
8. S. Lyaqini, M. Quafafou, M. Nachaoui, A. Chakib, Supervised learning as an inverse problem based on non-smooth loss function. Knowl. Inform. Syst., 1–20 (2020)
9. A. Défossez, L. Bottou, F. Bach, N. Usunier, On the convergence of adam and adagrad, arXiv preprint arXiv:2003.02395
10. Z. Fei, Z. Wu, Y. Xiao, J. Ma, W. He, A new short-arc fitting method with high precision using adam optimization algorithm. Optik **212**, 164788 (2020)
11. L. Bottou, *Stochastic Gradient Descent Tricks* (Springer, Berlin, Heidelberg, 2012), pp. 421–436
12. S. Lyaqini, M. Nachaoui, M. Quafafou, Non-smooth classification model based on new smoothing technique. J. Phys.: Confer. Seri. **1743**, IOP Publishing, p. 012025 (2021)
13. M. Nachaoui, Parameter learning for combined first and second order total variation for image reconstruction, Advanced Mathematical Models & Applications 5 (1) (2020) 53–69
14. https://archive.ics.uci.edu/ml/datasets/banknote+authentication

Identification of Robin Coefficient in Elliptic Problem by a Coupled Complex Boundary Method

L. Afraites and A. Oulmelk

Abstract In this paper, we study a Robin inverse problem governed by an elliptic partial differential equations. The unknown Robin coefficient is to be determined by additional boundary measurements, including both Dirichlet and Neumann boundary conditions. To solve this problem, we apply the coupled complex boundary method (CCBM) originally proposed in [4]. With CCBM, we introduce a complex elliptic partial differential equation with a boundary condition coupling the Dirichlet and Neumann boundary data, and we optimize the imaginary part of the solution in the domain to determine the Robin coefficient. Afterward, we use Tikhonov regularization to obtain a stable approximate Robin coefficient where we compute the gradient and the second derivative of our cost function. The latter expression allows us to show the existence of the optimum. Numerically, we use the gradient method and the finite element method for discretization. Finally, we give some numerical examples which show the feasibility and the stability of the proposed method.

Keywords Inverse Robin problem · Coupled method · Adjoint state

1 Introduction

This paper deals with the problem of identification a Robin coefficient in a stationary diffusion equation from boundary measurements, this coefficient is modeled the convection between the conducting body and the ambient environment, and its values are of significant practical interest in thermal and diffusion problems [10–12]. To fix ideas, let D be a known smooth and bounded domain in $\mathbb{R}^2$ or $\mathbb{R}^3$. Assume that ω is also an known but connected sub-domain of D such that $\partial\omega \cap \partial D = \emptyset$. In the following, we pose $\Gamma := \partial\omega$, $\Sigma := \partial\Omega$ and $\Omega = D \setminus \omega$. Once for a given Dirichlet boundary condition f, the value of the Neumann data is known to be g, then the electrostatic potential u solves the overdetermined problem.

L. Afraites (✉) · A. Oulmelk
FST-USMS, Beni-Mellal, Morocco

A. Nachaoui et al. (eds.), *Mathematical Control and Numerical Applications*,
Springer Proceedings in Mathematics & Statistics 372,
https://doi.org/10.1007/978-3-030-83442-5_6

$$\begin{cases} -\Delta u = 0 \ in \ \Omega, \\ u = f \ on \ \Sigma, \\ \partial_n u = g \ on \ \Sigma, \\ \partial_n u + qu = 0 \ on \ \Gamma, \end{cases} \tag{1}$$

where $f \in H^{\frac{1}{2}}(\Sigma)$ and $g \in (H^{\frac{-1}{2}}(\Sigma))$.

The Robin inverse problem is presented by:

Problem 1 Given f and g in Σ, find q such that the solution of the boundary value problem

$$\begin{cases} -\Delta u = 0 \text{ in } \Omega, \\ \partial_n u = g \text{ on } \Sigma, \\ \partial_n u + qu = 0 \text{ on } \Gamma, \end{cases} \tag{2}$$

satisfies:

$$u = f \text{ on } \Sigma.$$

Extensive literature exists for theoretical and numerical investigation of the Robin coefficient identification problem. For instance, Bangti and Jun [8] in his study of Robin inverse problem proposed a finite-element method to estimate the Robin coefficient in a stationary diffusion equation, Chantasiriwan [3] investigated the sensitivity of the Robin coefficient with respect to experimental uncertainties using the boundary element method, Chaabane [2] studied the stability of the Kohn and Vogelius method for the Robin inverse problem, Jin [9] applied the Conjugate Gradient Method (CGM) for the Robin inverse problem associated with the Laplace equation.

In general, the classical way to solve this problem, is to transform it in parameters optimization problem where the unknown is the Robin coefficient q. For this, there are two cost functions proposed, the first is the least square defined by [8]:

$$J_{LS}(q) = \frac{1}{2}\int_{\Sigma} |u - f|^2 d\sigma \tag{3}$$

where u solve the problem:

$$\begin{cases} -\Delta u = 0 \text{ in } \Omega, \\ \partial_n u = g \text{ on } \Sigma, \\ \partial_n u + qu = 0 \text{ on } \Gamma. \end{cases} \tag{4}$$

The second cost function is Khon-Vogeliu defined by [2]

$$J_{KV}(q) = \int_{\Omega} |\nabla(u_D - u_N)|^2 dx + \int_{\Gamma} q|(u_D - u_N)|^2 d\sigma \tag{5}$$

where u_D solution of Dirichlet problem:

$$\begin{cases} -\Delta u_D = 0 \text{ in } \Omega, \\ u_D = f \text{ on } \Sigma, \\ \partial_n u_D + q u_D = 0 \text{ on } \Gamma. \end{cases} \tag{6}$$

and u_N solution of Neumman problem:

$$\begin{cases} -\Delta u_N = 0 \text{ in } \Omega, \\ \partial_n u_N = g \text{ on } \Sigma, \\ \partial_n u_N + q u_N = 0 \text{ on } \Gamma. \end{cases} \tag{7}$$

We note from the above statements that the minimization problems (3) and (5) use the Neumann data g and the Dirichlet data f sequentially.

In this paper, we propose a new coupled complex boundary method (CCBM) that uses both g and f data in a single PDE. The idea of the CCBM is to couple the Neumann data and Dirichlet data in a Robin boundary condition in such a way that the Neumann data and Dirichlet data are the real part and imaginary part of the Robin boundary condition, respectively. Consequently, the data needed to fit are transferred from boundary to interior domain. The coupled complex boundary method (CCBM) is first proposed by Cheng et al. in [4] for an inverse source problem, Rongfang et al. in [7] applied it to an inverse conductivity problem with one measurement and recently, is applied to parameter identification in elliptic problems by Cheng et al. in [13]. To the best of our knowledge, in the literature, it is the first time that the idea of the coupled complex boundary condition has been explored for solving inverse Robin problem.

The content of this paper is distributed into four more sections, and after the above introduction, the work is organized as follows. In details. Section 2, We reformulated the inverse problem with CCBM and give some theoretical results on the complex bounded value problem. In Sect. 3, We use the Tikhonov regularization to have a stable approximate solution, and we provide some theoretical analyses of the new regularization framework. Section 4 is devoted to the some numerical results to show the feasibility and efficiency of the new method.

2 A Reformulation with CCBM

We first introduce some notation for function spaces and assumptions on the data. For a set G (e.g. D, Ω, ω, Γ, Σ), we denote by $W^{m,s}(G)$ the standard real Sobolev space with the norm $\|.\|_{m,s,G}$, $W^{0,s}(G) = L^s(G)$. In particular, $H^m(G)$ represents $W^{m,2}(G)$ with the corresponding inner product $(.,.)_{m,G}$ and norm $\|.\|_{m,G}$. Also, let $\mathbf{H}^m(G)$ be the complex version of $H^m(G)$ with the inner product $((.,.))_{m,G}$ and norm $|||.|||_{m,G}$ defined as follows:

$$((u, v))_{m,G} = (u, \bar{v})_{m,G}, \quad |||v|||^2_{m,G} = ((v, v))_{m,G} \quad \forall u, v \in \mathbf{H}^m(G),$$

where $\bar{v}$ is the conjugate complex of v. Denote $V = H^1(\Omega)$, $\mathbf{V} = \mathbf{H}^1(\Omega)$, and $Q = L^2(\Omega)$, $\mathbf{Q} = \mathbf{L}^2(\Omega)$. In addition, the admissible set $\mathcal{A}$ of the Robin coefficient is given by

$$\mathcal{A} = \left\{q \in L^\infty(\Gamma);\ c_0 \leq q(x) \leq c_1\ \ a.e\ \ x \in \Gamma\right\}.$$

We consider the complex transformation of problem (1):

$$\begin{cases} -\Delta u = 0 & \text{in } \Omega, \\ \partial_n u + iu = g + if & \text{on } \Sigma, \\ \partial_n u + qu = 0 & \text{on } \Gamma. \end{cases} \tag{8}$$

where $i = \sqrt{-1}$ is imaginary unit. Assume that $u = u_1 + iu_2$ is a weak solution of (8). Then real-valued functions u_1, u_2 satisfy:

$$\begin{cases} -\Delta u_1 = 0 \text{ in } \Omega, \\ \partial_n u_1 - u_2 = g \text{ on } \Sigma, \\ \partial_n u_1 + qu_1 = 0 \text{ on } \Gamma. \end{cases} \tag{9}$$

$$\begin{cases} -\Delta u_2 = 0 \text{ in } \Omega, \\ \partial_n u_2 + u_1 = f \text{ on } \Sigma, \\ \partial_n u_2 + qu_2 = 0 \text{ on } \Gamma. \end{cases} \tag{10}$$

If $u_2 = 0$ in Ω and u_2 is harmonic, then by the continuity $u_2 = 0$ on Σ and $u_2 = 0$ on Γ. By green formula and the above properties, we have $\partial_n u_2 = 0$ on Σ. As a result, from PDE (10) and (9), (u_1, q) is a solution of the original problem (1). Conversely, if (u, q) is a solution of the original problem (1), then immediately, it satisfies (8). We conclude from the above discussion that the original inverse Robin problem is equivalent to the following problem.

Problem 2 Find $q \in U_{ad}$ such that

$$u_2 = 0 \ \ in \ \ \Omega,$$

where $u = u_1 + iu_2$ solves

$$\begin{cases} -\Delta u = 0 & \text{in } \Omega, \\ \partial_n u + iu = g + if & \text{on } \Sigma, \\ \partial_n u + qu = 0 & \text{on } \Gamma. \end{cases} \tag{11}$$

Before analyzing and studying the inverse Problem 2, we show a well-posedness result concerning the boundary value direct problem (11) and we give the first and second differentiability of the state which respact the coefficient q.

For any $u, v \in \mathbf{V}$, we define

$$a(q,u,v) = \int_\Omega \nabla u \nabla \bar{v} dx + i \int_\Sigma u\bar{v} d\sigma + \int_\Gamma q u \bar{v} d\sigma. \tag{12}$$

$$l(v) = \int_\Sigma g\bar{v} d\sigma + i \int_\Sigma f\bar{v} d\sigma. \tag{13}$$

Then the variational problem of the boundary value problem (11) is:

$$Find \quad u \in \mathbf{V} \quad such \quad that \quad a(q,u,v) = l(v) \qquad \forall v \in \mathbf{V} \tag{14}$$

To study our problem. We will need the folowing norm equivalence lemma

Lemma 1 *The norm* $|||u|||_{1,\Gamma}$ *defined by*

$$|||u|||^2_{1,\Gamma} = |||\nabla u|||^2_{0,\Omega} + |||u|||^2_{0,\Gamma}$$

is equivalent to the norm $|||.|||_{1,\Omega}$ *on* $\mathbf{H}^1(\Omega)$.

The existence and uniqueness of the variational problem is given by the following Lemma.

Lemma 2 *For given* $f \in H^{\frac{1}{2}}(\Sigma)$ *and* $g \in (H^{\frac{1}{2}}(\Sigma))'$, *there existe a unique weak solution* $u \in \mathbf{V}$ *for problem (14).*

Proof We apply the complex version of the Lax-Milgram lemma. [5] by showing that the antilinear form $a(q,.,.)$ is continuous and elliptic on $\mathbf{V}$, and the linear form $l(.)$ is continuous on $\mathbf{V}$.
Ellipticity of $a(.,.)$ is immediate:

$$\begin{aligned} \Re ea(q,v,v) &= \int_\Omega |\nabla v_1|^2 + |\nabla v_2|^2 \, dx + \int_\Gamma q\left(v_1^2 + v_2^2\right) dx \\ &\geq \min(1, c_0)\left(|||\nabla v|||^2_{0,\Omega} + |||v|||^2_{0,\Gamma}\right) \\ &\geq \alpha |||v|||^2_{1,\Gamma} \end{aligned} \tag{15}$$

where $\Re e$ is the real part of $a(q,v,v)$.
For any $u = u_1 + iu_2, v = v_1 + iv_2 \in \mathbf{V}$, we have

$$\begin{aligned} |\, a(q,u,v) \,|^2 = & \left[\int_\Omega \nabla u_1 \nabla v_1 + \nabla u_2 \nabla v_2 dx + \int_\Gamma q(u_1v_1 + u_2v_2)dx + \int_\Sigma (u_1v_2 - u_2v_1)dx\right]^2 \\ & + \left[\int_\Omega \nabla u_2 \nabla v_1 - \nabla u_1 \nabla v_2 dx + \int_\Gamma q(u_2v_1 - u_1v_2)dx + \int_\Sigma (u_1v_1 + u_1v_1)dx\right]^2. \end{aligned}$$

Now Cauchy-Schwarz inequality, the trace inequality imply that

$$|\, a(q,u,v) \,| \leq \beta |||u|||_{1,\Gamma} |||v|||_{1,\Gamma}$$

Appealing to Lemma 1 again gives that $a(.,.)$ is continuous.

Similarly, we have the continuity of the linear form $l(.)$:

$$| l(v) | \leq c\gamma |||v|||_{1,\Gamma}, \tag{16}$$

where $\gamma = \|f\|_{H^{\frac{1}{2}}(\Sigma)} + \|g\|_{(H^{\frac{1}{2}}(\Sigma))'}$, α and β are constants which are independent of q. Therefore, the variational problem(14) admits a unique solution $u \in \mathbf{V}$.

From the proof of Lemma 2, for each $q \in \mathcal{A}$, there is a unique solution $u \in \mathbf{V}$, we can then conclude the following estimate

$$|||u|||_{1,\Gamma} \leq \frac{c\gamma}{\alpha}, \tag{17}$$

where we have used $\mathfrak{Rea}(q, u, v) \leq| a(q, u, v) |$.

We therefore define a functional S from $\mathcal{A}$ to $\mathbf{V}$: $q \in \mathcal{A} \longmapsto S(q) = u \in \mathbf{V}$, u is the solution of variational problem (14). In the following proposition, we give the continuity of the operator S.

Proposition 1 *The operator S is continuous in $\mathcal{A}$ and we have:*

$$|||S(q_1) - S(q_2)|||_{1,\Gamma} \leq \frac{c\gamma}{\alpha^2} \|q_1 - q_2\|_{\infty,\Gamma} \tag{18}$$

Proof Let $u_1 = S(q_1)$ and $u_2 = S(q_2)$, It is easy to verify that $a(q_1, u_1, v) = l(v) = a(q_2, u_2, v)$ for all $v \in \mathbf{V}$. By taking $w = u_1 - u_2$, then w satisfies the following equation

$$\begin{cases} -\Delta w = 0 & \text{in } \Omega, \\ \partial_n w + iw = 0 & \text{on } \Sigma, \\ \partial_n w + q_1 w = -(q_1 - q_2)u_2 & \text{on } \Gamma, \end{cases} \tag{19}$$

which gives

$$a(q_1, w, v) = -\left((q_1 - q_2)u_2, \bar{v}\right)_{0,\Gamma} \qquad \forall v \in \mathbf{V}. \tag{20}$$

Set $v = w$ and using (15) as well as Cauchy-Schwarz inequality to yield

$$\alpha |||w|||_{1,\Gamma}^2 \leq \|q_1 - q_2\|_{\infty,\Gamma} |||u_2|||_{0,\Gamma} |||w|||_{0,\Gamma} \tag{21}$$

From Lemma 1 and (17), we have

$$|||w|||_{1,\Gamma} \leq \frac{1}{\alpha} \|q_1 - q_2\|_{\infty,\Gamma} |||u_2|||_{1,\Gamma} \leq \frac{c\gamma}{\alpha^2} \|q_1 - q_2\|_{\infty,\Gamma}.$$

This completes the proof of the proposition.

Given that q in $\mathcal{A}$, for all h sufficiently small such as $q + h \in \mathcal{A}$ and $S(q+h) \in \mathbf{V}$ is well-defined. Let $w^1 = S(q+h) - S(q)$, then by definition of S, w^1 satisfies the following equation

$$\begin{cases} -\Delta w^1 = 0 & \text{in } \Omega, \\ \partial_n w^1 + i w^1 = 0 & \text{on } \Sigma, \\ \partial_n w^1 + (q+h) w^1 = -hu & \text{on } \Gamma, \end{cases} \tag{22}$$

which implies

$$a(q+h, w^1, v) = -(hu, \bar{v})_{0,\Gamma} \quad \forall v \in \mathbf{V} \tag{23}$$

Now, we can study the differentiability of S, and deduce the form of $DS(q)$.

Proposition 2 *For any $q \in \mathcal{A}$ the function $S(q)$ is differentiable with respect to q, and $u' = DS(q)h$ is the unique solution to the following sensitivity equation*

$$\begin{cases} -\Delta u' = 0 & \text{in } \Omega, \\ \partial_n u' + i u' = 0 & \text{on } \Sigma, \\ \partial_n u' + q u' = -hu & \text{on } \Gamma, \end{cases} \tag{24}$$

his variational equation

$$a(q, u', v) = -(hu, \bar{v})_{0,\Gamma} \quad \forall v \in V \tag{25}$$

with $u = S(q)$. Moreover,

$$|||DS(q)|||_{1,\Gamma} \leq \frac{c\gamma}{\alpha^2}. \tag{26}$$

Proof It is easy to verify that (25) has a unique solution $u' \subset \mathbf{V}$. Combine (23) and (25) to get

$$a(q, w^1 - u', v) = -\left(hw^1, \bar{v}\right)_{0,\Gamma} \quad \forall v \in \mathbf{V}.$$

By taking $v = w^1 - u'$, we obtain

$$a(q, w^1 - u', w^1 - u') = -\left(hw^1, \overline{w^1} - \overline{u'}\right)_{0,\Gamma}.$$

we apply ellipticity (15) of $a(q, ., .)$ and Cauchy-Schwarz inequality, we obtain

$$\alpha|||w^1 - u'|||_{1,\Gamma}^2 \leq \|h\|_{\infty,\Gamma}|||w^1|||_{0,\Gamma}|||w^1 - u'|||_{0,\Gamma}.$$

From Lemma 1, we have

$$\alpha|||w^1 - u'|||_{1,\Gamma}^2 \leq \|h\|_{\infty,\Gamma}|||w^1|||_{1,\Gamma}|||w^1 - u'|||_{1,\Gamma},$$

which gives

$$|||w^1 - u'|||_{1,\Gamma} \leq \frac{1}{\alpha}\|h\|_{\infty,\Gamma}|||w^1|||_{1,\Gamma}. \tag{27}$$

Appealing to Proposition 1 again for (22), we obtain:

$$|||w^1|||_{1,\Gamma} \leq \frac{c\gamma}{\alpha^2}\|h\|_{\infty,\Gamma}. \tag{28}$$

From (27) and (28), we have

$$|||w^1 - u'|||_{1,\Gamma} \leq \frac{c\gamma}{\alpha^3}\|h\|^2_{\infty,\Gamma}. \tag{29}$$

which gives

$$\frac{|||S(q+h) - S(q) - u'|||_{1,\Gamma}}{\|h\|_{\infty,\Gamma}} = \frac{|||w^1 - u'|||_{1,\Gamma}}{\|h\|_{\infty,\Gamma}} \leq \frac{c\gamma}{\alpha^3}\|h\|_{\infty,\Gamma}. \tag{30}$$

Thus, S is differentiable at q and $DS(q)h = u'$.
Also, let $v = u'$ in (25), and use ellipticity (15) of $a(q,.,.)$ as well as Cauchy-Schwarz inequality again to obtain

$$\alpha|||u'|||^2_{1,\Gamma} \leq \|h\|_{\infty,\Gamma}|||u|||_{0,\Gamma}|||u'|||_{0,\Gamma}.$$

From Lemma 1 and (17), we have

$$|||u'|||_{1,\Gamma} \leq \frac{c\gamma}{\alpha^2}\|h\|_{\infty,\Gamma}. \tag{31}$$

which shows that $DS(q)$ is uniformly bounded.

Among the objectives of this work is to calculate the second derivative of S. For this purpose, denoted by $w^2 = DS(q+h_1)h_2 - DS(q)h_2$, where $DS(q+h_1)h_2$ and $DS(q)h_2$ are the derivatives of function S at points $q+h_1$ and q, respectively. By Proposition 2, we have
$u'(q+h_1) = DS(q+h_1)h_2$ satisfies the following equation

$$\begin{cases} -\Delta u'(q+h_1) = 0 & \text{in } \Omega, \\ \partial_n u'(q+h_1) + iu'(q+h_1) = 0 & \text{on } \Sigma, \\ \partial_n u'(q+h_1) + (q+h_1)u'(q+h_1) = -h_2 u(q+h_1) & \text{on } \Gamma, \end{cases} \tag{32}$$

where $u(q+h_1) = S(q+h_1)$ and $u'(p) = DS(q)h_2$ satisfies the following equation

$$\begin{cases} -\Delta u'(q) = 0 & \text{in } \Omega, \\ \partial_n u'(q) + iu'(q) = 0 & \text{on } \Sigma, \\ \partial_n u'(q) + qu'(q) = -h_2 u(q) & \text{on } \Gamma, \end{cases} \tag{33}$$

where $u(q) = S(q)$.

Therefore, with preliminary computation, w^2 satisfies the following equation

$$\begin{cases} -\Delta w^2 = 0 & \text{in } \Omega, \\ \partial_n w^2 + i w^2 = 0 & \text{on } \Sigma, \\ \partial_n w^2 + (q+h_1)w^2 = -h_2 u(q+h_1) + h_2 u(q) - h_1 u'(q) & \text{on } \Gamma, \end{cases} \tag{34}$$

his variational equation

$$a(q+h_1, w^2, v) = -\left(h_2\left[S(q+h_1) - S(q)\right], \bar{v}\right)_{0,\Gamma} - \left(h_1 DS(q)h_2, \bar{v}\right)_{0,\Gamma} \quad \forall v \in \mathbf{V} \tag{35}$$

from which we have the following result.

Proposition 3 *For any $q \in \mathcal{A}$ the function $S(q)$ is twice-differentiable with respect to q, and $u'' = DS^2(q)(h_1, h_2)$ is the unique solution to the variational equation*

$$a(q, u'', v) = -\left(h_2 DS(q)h_1, \bar{v}\right)_{0,\Gamma} - \left(h_1 DS(q)h_2, \bar{v}\right)_{0,\Gamma} \quad \forall v \in \mathbf{V} \tag{36}$$

with $u = S(q)$. Moreover,

$$|||DS^2(q)|||_{1,\Gamma} \le \frac{c\gamma}{\alpha^2}. \tag{37}$$

Proof From the complex version of the Lax-Milgram lemma [5], the variational equation (36) has a unique solution $u'' \in \mathbf{V}$. Moreover, we Combine (35) and (36) to get

$$a(q+h_1, w^2, v) - a(q, u'', v) = -\left(h_2\left[S(q+h_1) - S(q) - DS(q)h_1\right], \bar{v}\right)_{0,\Gamma},$$

which implies

$$a(q, w^2 - u'', v) = -\left(h_2\left[S(q+h_1) - S(q) - DS(q)h_1\right], \bar{v}\right)_{0,\Gamma} - \left(h_1 w^2, \bar{v}\right)_{0,\Gamma}.$$

By taking $v = w^2 - u''$ and using ellipticity (15) of $a(q, ., .)$ and Cauchy-Schwarz inequality, we obtain

$$\alpha|||w^2 - u''|||_{1,\Gamma}^2 \le \|h_2\|_{\infty,\Gamma}|||S(q+h_1) - S(q) - DS(q)h_1|||_{0,\Gamma}|||w^2 - u''|||_{0,\Gamma} \tag{38}$$

$$+\|h_1\|_{\infty,\Gamma}|||w^2|||_{0,\Gamma}|||w^2 - u''|||_{0,\Gamma} \tag{39}$$

From Lemma 1, we have

$$|||w^2 - u''|||_{1,\Gamma} \le \frac{1}{\alpha}\|h_2\|_{\infty,\Gamma}|||S(q+h_1) - S(q) - DS(q)h_1|||_{1,\Gamma} + \frac{1}{\alpha}\|h_1\|_{\infty,\Gamma}|||w^2|||_{1,\Gamma} \tag{40}$$

From (30) and (31), with similar arguments, we have

$$|||S(q+h_1) - S(q) - DS(q)h_1|||_{1,\Gamma} \leq \frac{c\gamma}{\alpha^3}\|h_1\|^2_{\infty,\Gamma} \tag{41}$$

$$|||w^2|||_{1,\Gamma} \leq \frac{c\gamma}{\alpha^3}\|h_1\|_{\infty,\Gamma}\|h_2\|_{\infty,\Gamma} \tag{42}$$

Combining (40) and (41)–(42) we obtain

$$|||w^2 - u''|||_{1,\Gamma} \leq \frac{c\gamma}{\alpha^3}\|h_1\|^2_{\infty,\Gamma}\|h_2\|_{\infty,\Gamma}, \tag{43}$$

then

$$\frac{|||DS(q+h_1)h_2 - DS(q)h_2 - u''|||_{1,\Gamma}}{\|h_1\|_{\infty,\Gamma}} = \frac{|||w^2 - u''|||_{1,\Gamma}}{\|h_1\|_{\infty,\Gamma}} \leq \frac{c\gamma}{\alpha^3}\|h_1\|_{\infty,\Gamma}\|h_2\|_{\infty,\Gamma},$$

which shows that S is twice-differentiable at a and $u'' = DS^2(q)(h_1, h_2)$
Also, taking $v = u''$ in (36) and combine (26) to get

$$|||u''|||_{1,\Gamma} \leq \frac{c\gamma}{\alpha^3}\|h_1\|_{\infty,\Gamma}\|h_2\|_{\infty,\Gamma},$$

we concluded to (37).

3 Tikhonov Regularization

In the section, we give the Tikhonov regularization framework for inverse Problem(2), For this purpose,denote $\mathcal{A} = L^\infty(\Gamma)$ and $u := S(q) = u_1(q) + iu_2(q) \in \mathbf{V}, \ \forall q \in \mathcal{A}$ the weak solution of boundary value problem (11). Moreover, since $\mathcal{A}$ can be embedded into $L^2(\Gamma)$. From all that, we transform the our problem 2 into following regularized parameters optimization:

$$J(q) = \frac{1}{2}\|u_2(q)\|^2_{L^2(\Omega)} + \frac{\beta}{2}\|q\|^2_{L^2(\Gamma)} \tag{44}$$

where β is regularization parameter and we introduce the following minimization problem:

Problem 3 Find $\bar{q} \in \mathcal{U}_{ad}$ in admissible shape, such that:

$$\bar{q} = \min_{q \in \mathcal{U}_{ad}} J(q).$$

In the rest of our work, we propose $\mathcal{U}_{ad}$ is a finite dimensional closed convex subset of $\mathcal{A}$, for instance, $\mathcal{U}_{ad}$ can be the subset of $\mathcal{A}$ of constant functions or polynomials of some degree, corresponding to certain partition of Ω,

In that case, we can prove that the functional cost is convex and that its minimizer in $\mathcal{U}_{ad}$ is unique. For any $q \in \mathcal{U}_{ad}$, and $h \in \mathcal{U}_{ad}$, we have

$$J'(q)h \quad = \left(u_2(q), u_2'\right)_{0,\Omega} + \beta\,(q, h)_{0,\Gamma}\,, \tag{45}$$

$$J''(q)h = \|u_2'\|_{0,\Omega}^2 + \left(u_2(q), u_2''\right)_{0,\Omega} + \beta\|h\|_{0,\Gamma}^2, \tag{46}$$

where $u_2 = Im(u)$, $u_2' = Im(u')$, $u_2'' = Im(u'')$, $u' = DS(q)h$ is the unique solution to the variational equation (25) and $u'' = DS^2(q)(h, h)$ is the unique solution to the variational equation (36).

The result of convexity of the functional cost is established in the following proposition

Proposition 4 *For any $\beta > 0$, there exists a positive constant d independent of $q \in \mathcal{A}$ such that for $d < \beta$, the functional $J(q)$ is strictly convex.*

Proof It is easy to see that the first term and the third term in $J''(q)h$ are non-nigative, it remains to show that the second term in $J''(q)h$ is positive,

From the Cauchy-Schwarz inequality, Lemma 1 and (37), we have

$$\left(u_2(q), u_2''\right)_{0,\Omega} \le \|u_2\|_{0,\Omega}\|u_2''\|_{0,\Omega} \tag{47}$$

$$\le \|u_2\|_{0,\Omega}\||u''\||_{1,\Gamma} \tag{48}$$

$$\le \tfrac{c\gamma}{\alpha^3}\|u_2\|_{0,\Omega}\|h\|_{\infty,\Gamma}^2 \tag{49}$$

From (17), we have

$$\|u_1\|_{0,\Omega} \le \||u\||_{0,\Omega} \le \||u\||_{1,\Gamma} \le \frac{c\gamma}{\alpha} := M.$$

By Eq. (10), we obtain

$$\|u_2\|_{0,\Omega} \qquad \le \|u_2\|_{1,\Omega} \tag{50}$$

$$\le \tfrac{c}{\alpha}\left(\|f\|_{-\frac{1}{2},\Sigma} + \|u_1\|_{-\frac{1}{2},\Omega}\right) \tag{51}$$

$$\le \tfrac{c}{\alpha}\left(\|f\|_{-\frac{1}{2},\Sigma} + \|u_1\|_{0,\Omega}\right) \tag{52}$$

$$\le \tfrac{c}{\alpha}\left(\|f\|_{-\frac{1}{2},\Sigma} + M\right) := N \tag{53}$$

Thus

$$|\left(u_2(q), u_2''\right)_{0,\Omega}| \le \frac{c\gamma N}{\alpha^3}\|h\|_{\infty,\Gamma}^2 := d\|h\|_{\infty,\Gamma}^2$$

then

$$J''(q)h \ge \|u_2'\|_{0,\Omega}^2 + (\beta - d)\|h\|_{0,\Gamma}^2$$

If we let $d < \beta$, then the sum of the second and third term of $J''(q)h$ is strictly positive, and $J''(q)h > 0$, thus $J(q)$ is strictly convex.

Now, we give a well-posedness result and the first optimality condition of the solution of Problem 3 as follows:

Proposition 5 *For any $\beta > 0$, let $d < \beta$, then Problem 3 has a unique solution $q \in \mathcal{U}_{ad}$ which depends continuously on all data. Moreover, q is characterized by the inequality*

$$(v_2 u_1 - v_1 u_2 + \beta q, p - q)_{0,\Gamma} \geq 0, \tag{54}$$

where $v = v_1 + i v_2 \in \mathbf{V}$ is the unique weak solution of the adjoint problem:

$$\begin{cases} -\Delta v = u_2 \text{ in } \Omega, \\ \partial_n v - i v = 0 \text{ on } \Sigma, \\ \partial_n v + q v = 0 \text{ on } \Gamma. \end{cases} \tag{55}$$

and $u = S(q) = u_1 + i u_2 \in \mathbf{V}$ is the weak solution of the boundary value problem (11).

Proof By the assumptions, we have $\mathcal{U}_{ad}$ is a closed and convex set of Hilbert space A and J is functional strictly convex. Then, by using a standard result on convex minimization problems [1, 6], there is a unique stable solution $q \in \mathcal{U}_{ad}$ to Problem 3 and the solution is characterized by the optimality condition

$$J'(q)(p - q) \geq 0 \;\; \forall p \in \mathcal{U}_{ad}. \tag{56}$$

To show (54), it suffices to calculate the gradient of the functional J. By taking $h = p - q$ in (56), we obtain

$$J'(q)h \geq 0 \;\; \forall p \in \mathcal{U}_{ad}. \tag{57}$$

From (45), we have

$$\left(u_2(q), u_2'\right)_{0,\Omega} + \beta\,(q, h)_{0,\Gamma} \geq 0, \tag{58}$$

where $u' = DS(q)h \in \mathbf{V}$ is a solution of the sensitivity problem (24).

Multiplying the adjoint function $v \in \mathbf{V}$ on both sides of the first equation in (24), the other side, multiplying the function $u' \in \mathbf{V}$ on both sides of the first equation in (55), integrating over Ω and using the Green formula and the boundary conditions of the two probems, we obtain

$$\begin{cases} \displaystyle\int_\Omega \nabla u' \nabla \bar{v} dx + \int_\Sigma i u' \bar{v} dx + \int_\Gamma q u' \bar{v} dx = -\int_\Gamma h u \bar{v} dx \quad \forall v \in \mathbf{V} \\ \displaystyle\int_\Omega \nabla v \nabla \bar{u}' dx - \int_\Sigma i v \bar{u}' dx + \int_\Gamma q v \bar{u}' dx = \int_\Omega u_2 \bar{u}' dx \quad \forall u' \in \mathbf{V} \end{cases} \tag{59}$$

We take the imaginary part to have the term $\left(u_2(q), u_2'\right)_{0,\Omega}$, we have

$$\begin{cases} \int_\Omega (\nabla u_2' \nabla v_1 - \nabla u_1' \nabla v_2)dx + \int_\Sigma (u_1' v_1 + u_2' v_2)dx + \int_\Gamma q(u_2' v_1 - u_1' v_2)dx = -\int_\Gamma h(u_2 v_1 - u_1 v_2)dx \\ \int_\Omega (\nabla v_2 \nabla u_1' - \nabla v_1 \nabla u_2')dx - \int_\Sigma (u_1' v_1 + u_2' v_2)dx + \int_\Gamma q(v_2 u_1' - v_1 u_2')dx = -\int_\Omega u_2 u_2' dx \end{cases}$$

we combine the resulting relations in (60) to get

$$\left(u_2(q), u_2'\right)_{0,\Omega} = (v_2 u_1 - v_1 u_2, h)_{0.\Gamma}, \tag{60}$$

this completes the proof of the proposition.

4 Numerical Results

In this section, we present some numerical simulations in order to confirm and complete our previous theoretical results. We present the gradient algorithm combined the element finite method to solve the direct and adjoint problem. The gradient algorithm is presented in the following Algorithm.

Algorithm 1 Gradient algorithm for inverse Robin coefficient

1. Choose an initial coefficient q_0, the regularization parameter β and set $k = 0$.
2. Solve the state problem (11).
3. Solve the adjoint problem (55).
4. Compute the gradient by the expression (54)
5. Update the current coefficient Robin : $q_{k+1} = q_k - \alpha J'(q_k)$, for some sufficiently small scalar $\alpha > 0$.
6. While $\|J'(q_k)\| \geq \epsilon$, $k = k + 1$ and go back to steep 2 and repeat .

The numerical simulations presented are made in dimension two using the finite elements method for the direct (11) and the adjoint problem (55). We consider $\Omega = (0, 1) \times (0, 1)$, the part of the boundary Γ is taken to be $\Gamma = (0, 1) \times \{0\}$ and the other part of the boundary Σ we take it as follows $\Sigma = \partial\Omega \backslash \Gamma$.

We construct the synthetic data f by fixing the unknown Robin coefficient and choosing the Neumann condition g, then we solve the Neumann problem 7 using the finite elements method and we extract the additive additional data f as the solution on the boundary Σ. In our test cases the simulated noisy data are generated using the following formula:

$$\tilde{f} = f + \varepsilon f.(2.rand(size(f) + 1)$$

Example 1 Let $u(x_1, x_2) = x_1^2 - x_2^2$ the exact solution for the Neumann problem 7 and the function f is obtained by solving the Neumann problem 7, and the Robin coefficient $q(x)$ will be estimate is given by

$$q(x_1) = (1 + x_1^2)$$

Example 2 The second example is synthetic, we consider g as normal derivative in above example and we suppose the unknown Robin coefficient as follows:

$$q(x_1) = \frac{1}{4} \cos^2(2\pi x_1). \tag{61}$$

The measurement f is obtained by solving the Neumann problem (7).

Example 3 The third example is synthetic also with same data g and we suppose the unknown Robin coefficient as follows:

$$q(x_1) = e^{-2x_1}. \tag{62}$$

The Dirichlet data f is obtained by solving the Neumann problem (7).

4.1 Identification Results Without Noise

Now we present the results of the estimate without noise the Robin coefficient q for the previous three examples. Figure 1 shows the numerical results of Robin coefficient reconstruction compared to exact one and the evolution of the cost functional and gradient with respect to number of iteration from Exemple 1. In Fig. 2, we present the numercial results obtained by our methoud for Example 2 and Example 3 without noise.

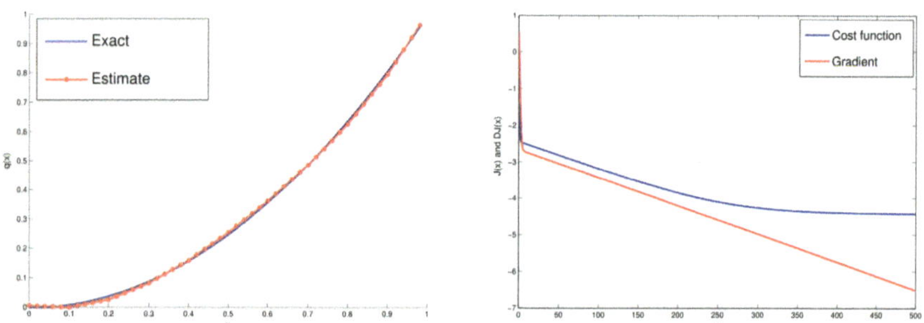

(a) Example 2 without noise (b) The cost function and the gradient

Fig. 1 The numerical results of descent algorithm for Example 1 without noise

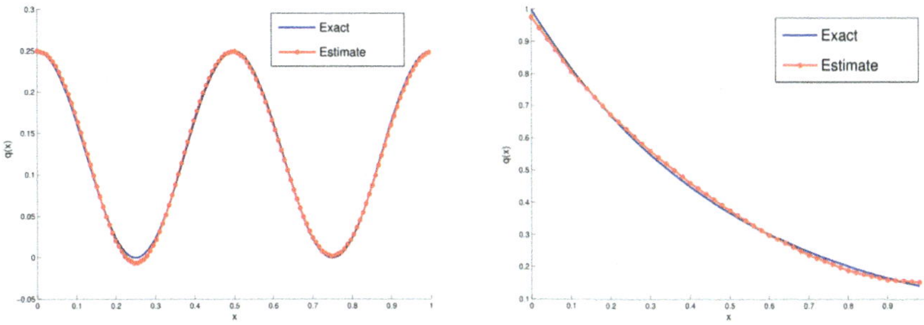

(a) Example 2 without noise (b) Example 3 whitout noise

Fig. 2 The numerical results of descent algorithm for Examples 2 and 3 without noise

4.2 Identification Results with Noise

In this subsection, we shows the numerical results obtained by noisy data. Figure 3 shows the numerical results and the evolution of the cost functional and their gradient with a noise with respect the number of iterations. In Fig. 4, we present the numercial results obtained by our methoud for Example 2 and Example 3 with noise.

In the all tests, the results obtained show the advantage of the our approach in terms of reconstruction quality and in terms of the evolution of the cost function and gradient with respect to the number of iterations.

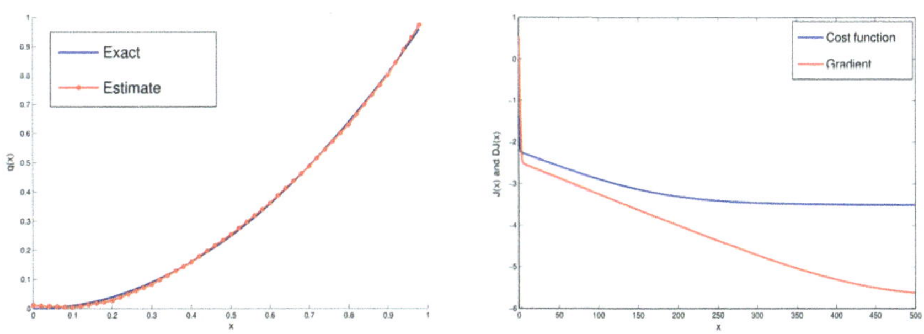

(a) Example 1 with 5% of noise (b) the cost function and the gradient

Fig. 3 The numerical results of descent algorithm for Example 1 with noise

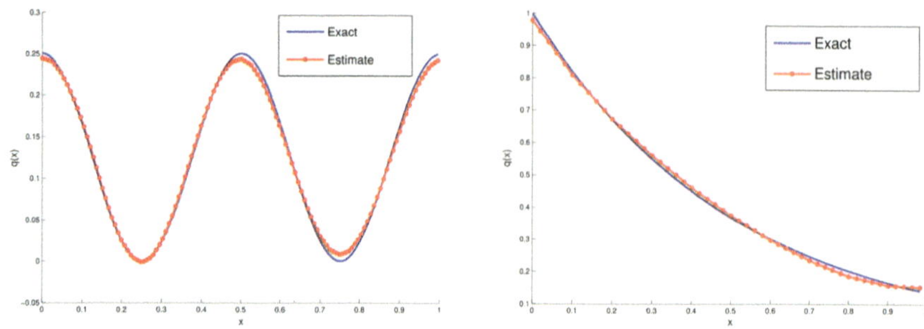

(a) Example 2 with 5% of noise (b) Example 3 with 5% of noise

Fig. 4 The numerical results of descent algorithm for Examples 2 and 3 with noise

References

1. K. Atkinson, W. Han, *Theoretical Numerical Analysis: A Functional Analysis Framework*, 3rd edn. (Springer, New York, 2009)
2. S. Chaabane, C. Elhechmi, and M. Jaoua, A stable recovery method for the Robin inverse problem, Math. Comput. Simul., 66, 367–383, 2014
3. S. Chantasiriwan, Inverse determination of steady-state heat transfer coefficient, Int. Commun. Heat Mass Transf, 27, 1155–1164, 2000
4. X.L. Cheng, R.F. Gong, W. Han, X. Zheng, A novel coupled complex boundary method for solving inverse source problems. Inverse Probl. **30**, 055002 (2014)
5. R. Dautray, J.-L. Lions, *Mathematical Analysis and Numerical Methods for Science and Technology*, vol. 2 (Springer, Berlin, 1988)
6. K. Glashoff, S.A. Gustafson, *Linear Optimization and Approximation: An Introduction to the Theoretical Analysis and Numerical Treatment of Semi-Infinite Programs* (Springer, New York, 1983)
7. R. Gong, X. Cheng, W. Han, A coupled complex boundary method for an inverse conductivity problem with one measurement. Appl. Anal. Int. J. (2016)
8. B. Jin, J. Zou, Numerical estimation of the Robin coefficient in a stationary diffusion equation IMA. J. Numer. Anal. **30**, 677–701 (2010)
9. B. Jin, Conjugate gradient method for the Robin inverse problem associated with the Laplace equation, Int. J. Numer. Methods Eng., 71, 433–453, 2007
10. T. J. Martin and G. S. Dulikravich, Inverse determination of steady heat convection coefficient distribution J. Heat Transf., 120, 328–334, 1998
11. A. M. Osman and J. V. Beck, Nonlinear inverse problem for the estimation of time-and-space dependent heat transfer coefficients, J. Thermophys. Heat Transf, 3, 146–152, 1989
12. F.M. White, *Heat and Mass Transfer*. Reading, MA: Addison-Wesley
13. X. Zheng, X. Cheng, R. Gong, A coupled complex boundary method for parameter identification in elliptic problems. Int. J. Comput. Math. (2019)

A New Space-Variant Optimization Approach for Image Segmentation

M. Nachaoui, A. Laghrib, and M. Hakim

Abstract In this paper, we introduce a new variational image segmentation model based on a non-convex regularization and an efficient parameter selection to better conserve the image structures and control the segmentation of the boundary with less intensity inhomogeneities. The proposed regularization associated to the automatic selection of the parameter λ allows us to segment an image with low intensities loss. In addition, we introduce fast and efficient optimization algorithms that can handle the non-smoothness of the objective function. In fact, the output segmented regions in the image, as well as edges construction are better conserved. Selected numerical experiments demonstrate that the segmented images can cope with some sate of art segmentation results.

Keywords Image segmentation · Variational approach · Non-convex regularization · Parameter selection · ADMM algorithm

1 Introduction

Image segmentation is one of the most critical tasks of image processing. Its aim is extracting information from an image via segmentation, which have considerable influence over the accuracy of object representation and feature measurement. Concretely, image segmentation is the process of portioning an image into its representatives parts inclusind intensities, edges, color changes, and so on. Several investigation on image segmentation has been developed, many research directions have been exploited, some very different methodology have been adopted and a wide variety of segmentation algorithms have been proposed in the literature [9, 11, 17, 22, 23, 28]. Nevertheless, none of the proposed segmentation algorithms can be applicable to all images and different algorithms are not equally suitable for particular applications.

M. Nachaoui (✉) · A. Laghrib
LMA FST Béni-Mellal, Université Sultan Moulay Slimane, Mellal, Morocco

M. Hakim
LAMAI, FSTG Marrakech, Université Cadi Ayyad, Marrakesh, Morocco

A. Nachaoui et al. (eds.), *Mathematical Control and Numerical Applications*, Springer Proceedings in Mathematics & Statistics 372,
https://doi.org/10.1007/978-3-030-83442-5_7

Variational image segmentation approaches has been particularly studied over the past two decades [3, 18, 26]. In general, variational models consist of two terms, namely a fidelity term and a regularization term. As it's well known that the image segmentation is one of highly ill-posed problem. Therefore, regularization methods should be employed to ensure stable approximate solutions. This can be done by adding a regularization term [15, 16] or introducing some algorithms which has a regularizing effect [2, 5, 19, 20]. Among the theoretical and numerical requirement for a regularizer, its allows to the recovery of edges, and smooth the other homogeneous parts. Recently, various non-convex regularizers have been considered with better performance than the usual convex ones [10, 12]. Indeed, it permits to well preserve geometric shapes of the homogeneous regions, and to protect edges from over-smoothing which is a common drawback of the convex regularizer.

In this paper, we consider a new variational image segmentation model based on a non-convex regularization. In particular, we conduct an efficient parameter selection process. This allows better conservation of the image structures and good control of the segmentation over the boundary with less intensity inhomogeneities. Recently, the automatic selection of parameters by learning approach have increased [21, 24]. Unlike to the existent methods, the proposed regularization associated to the automatic selection of parameter allows us to segment an image with low intensities loss and without augmentation of computing complexity. In fact, we develope a new fast alternative iteration algorithm which combined ADMM iterative and automatic selection of regularzation parameter based on the estimation of the gradient variation in each image window [7]. Through comprehensive numerical experiments we demonstrate that the segmented regions in the image, as well as edges construction are better conserved. Moreover, the proposed approach achieves more accurate results in terms of visual evaluation and a variety of quantitative indices compared to the sate of art segmentation results.

The remains of this work is structured such as: In Sect. 2 we introduce the problem setting. Section 3 is devoted to the proposed ADMM algorithm. While the Sect. 4 presents some numerical experiments.

2 The Problem Formulation

In this work, knowing that $f \in \mathbb{R}^n$ is the initial image, we propose a space variant nonconvex variational segmentation model given by the $u \in \mathbb{R}^n$ solution of the following minimization problem

$$\min_u \left\{ \lambda ||u - f||_2^2 + \Psi(|\nabla u(x)|) \right\}, \tag{1}$$

where $\Psi(|\nabla u(x)|) = \int \phi(|\nabla u(x)|)\, dx$ and ϕ is a nonconvex function (presented in [10, 12]) defined by

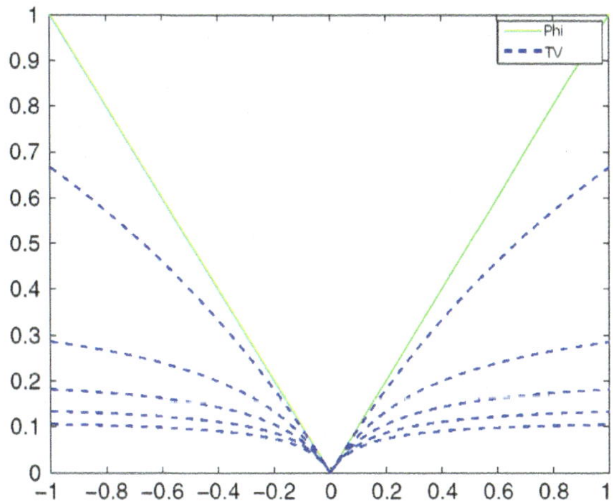

Fig. 1 Plot of the regularizer ψ with different ρ values $\rho \in \{0.5, 2.5, 4.5, 7, 8.5\}$

$$\phi(|t|) = \frac{|t|^p}{1+\rho|t|^p}, \text{ with: } p \geq 1. \tag{2}$$

ρ is the concavity parameter that control the non-convexity rate. In particular, we consider the non-smooth case where $p = 1$ and plot the regularizer ϕ with different values of ρ compared to the TV regularisation $\phi_{TV}(u) = \int |\nabla u(x)|\, dx$ in Fig. 1. With this choice, edges are protected from over-smoothing with minimal growth condition on ϕ. Hence, the regularizer ϕ does not penalize strong gradients of u, which will protect tiny edges in the segmented image.

The parameter $\lambda \in \mathbb{R}_+^*$ in Eq. (1) is a tuning parameter to keep balance between the first data fitting term and the regularization term. The choice of this parameter is very important, especially in the case where the fidelity term is non-convex since it is very sensitive to the image variations as shown in Fig. 2.

The parameter λ is in fact based on the estimation of the gradient variation in each image window [7]. In fact, if we plot the histogram of the image and its contour (see Fig. 3 for example), we can see that the histogram is normalized in the case of contours. To better recognize the edges in each region, we use then the spatial difference curvature approach [4], to discern edges in a given image. We first define the difference curvature coefficient $\mathfrak{C}$ for the ith pixel in the image by

$$\mathfrak{C}_i = \big||\mathfrak{s}_{\eta\eta}| - |\mathfrak{s}_{\epsilon\epsilon}|\big|, \tag{3}$$

where

$$\mathfrak{s}_{\eta\eta} = \frac{\mu_x^2\mu_{xx} + 2\mu_x\mu_y\mu_{xy} + \mu_y^2\mu_{yy}}{\mu_x^2 + \mu_y^2}, \tag{4}$$

and

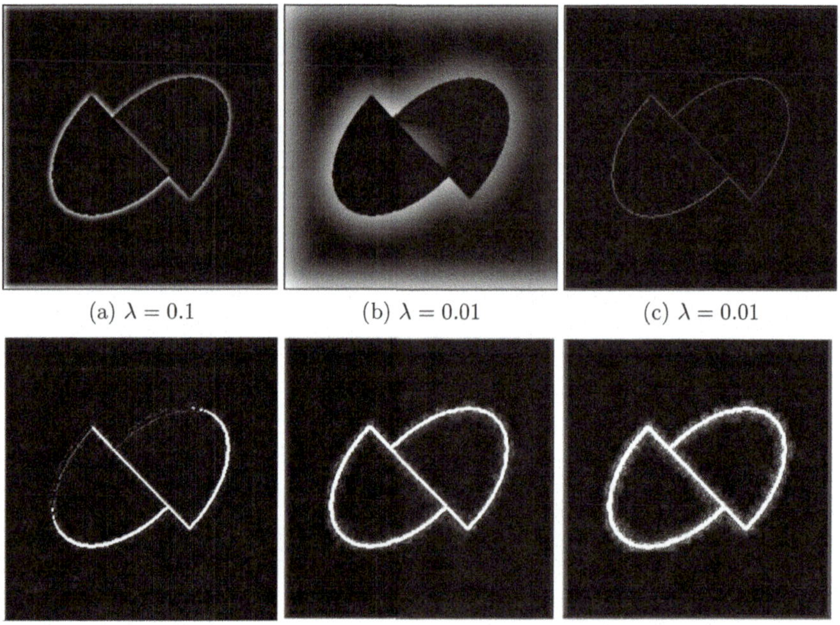

(a) $\lambda = 0.1$ (b) $\lambda = 0.01$ (c) $\lambda = 0.01$

(d) $\lambda = 0.5$ (e) $\lambda = 1$ (f) $\lambda = 10$

Fig. 2 The segmentation result of the (*Shepp-Logan phantom*) image with respect to different values of λ

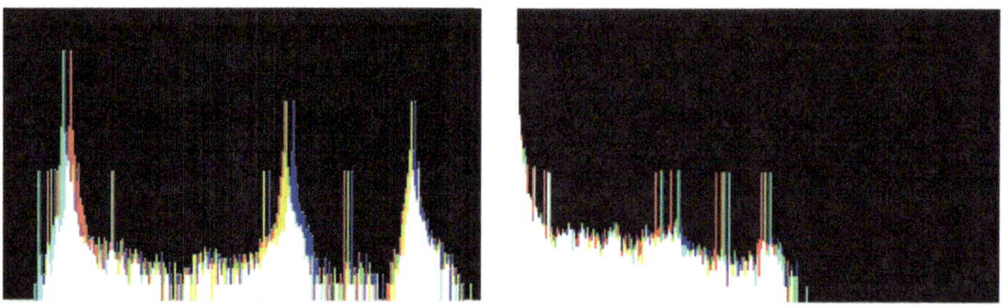

(a) u (b) $|\nabla u|$

Fig. 3 The Histogramme of the segmented (*Shepp-Logan phantom*) image compared to its contour

$$\mathfrak{s}_{\epsilon\epsilon} = \frac{\mu_y^2\mu_{xx} + 2\mu_x\mu_y\mu_{xy} + \mu_x^2\mu_{yy}}{\mu_x^2 + \mu_y^2}. \tag{5}$$

η and ϵ are respectively the direction of the gradient and the perpendicular direction to the gradient. μ_x, μ_y, μ_{xx}, μ_{yy} and μ_{xy} denote, respectively the first and second order derivative in each pixel. While $|\ |$ is the absolute value function. Thus, we use the following formula to approximate the parameter λ related to the appropriate balance between sharp edges and smooth ones.

$$\lambda = \begin{cases} \dfrac{1}{1+\kappa\,\mathrm{trace}(\mathfrak{W})} & \text{if } \|\nabla u\|_1 \leq \kappa \\ \exp(-\dfrac{\|\nabla u\|_1^2}{2}) & \text{if } \|\nabla u\|_1 > \kappa \end{cases}, \tag{6}$$

where κ is the maximum value of the grayscale image f. $\mathfrak{W}$ is a spatially weighted matrix defined by the difference curvature as follows

$$\mathfrak{W} = \begin{bmatrix} \mathfrak{C}_1 & 0 & \dots & 0 \\ 0 & \mathfrak{C}_2 & \dots & 0 \\ \vdots & \vdots & \ddots & \vdots \\ 0 & 0 & \dots & \mathfrak{C}_{n^2} \end{bmatrix}, \tag{7}$$

where 'trace' is the trace operator defined by $\mathrm{trace}(\mathfrak{W}) = \Sigma_i \mathfrak{C}_i$.

3 The ADMM Algorithm

In this section, we introduce the ADMM iterative algorithm [6, 8] used to solve the proposed segmentation model (1). For that we introduce the objective function, which is defined by

$$\mathfrak{L}(u; \lambda) = \lambda ||u - f||_2^2 + \Psi(|\nabla u(x)|, \tag{8}$$

The minimization problem (1) is now reformulated such as:

$$\min_u \mathfrak{L}(u; \lambda); \tag{9}$$

Since the functional $\mathfrak{L}$ is convex, we use the variable splitting technique [1] and insert the auxiliary variable v, such that the model (8) becomes a constrained linear problem with the following form:

$$\min_{u,v} \left\{\lambda ||u - f||_2^2 + \Psi(|v|)\right\} \quad \text{with: } v = \nabla u. \tag{10}$$

Now, to solve this minimization problem, we define the augmented Lagrangian functional L such as:

$$L(u, v, \xi) = \lambda ||u - f||_2^2 + \Psi(|v|) - < \xi, v - \nabla u > + \frac{\tau}{2} \|v - \nabla u\|_2^2, \tag{11}$$

where $\tau > 0$ is a scalar penalty parameter and $\xi \in \mathbb{R}^{2n}$ is the vector of Lagrange multipliers associated to the constrained minimization one (10). We then reformulate the following saddle-point problem:

$$\begin{aligned} \text{Find} \quad & (\widehat{u}, \widehat{v}, \widehat{\xi}) \in \mathbb{R}^n \times \mathbb{R}^{2n} \times \mathbb{R}^{2n} \\ \text{s.t.} \quad & L(\widehat{u}, \widehat{v}, \xi) \leq L(\widehat{u}, \widehat{v}, \widehat{\xi}) \leq L(u, v, \widehat{\xi}) \\ & \forall (u, v, \xi) \in \mathbb{R}^n \times \mathbb{R}^{2n} \times \mathbb{R}^{2n}. \end{aligned} \tag{12}$$

Assuming that the initial candidate is previously computed or initialized vectors v^{k-1} and ξ^k, the k-th iteration of the introduced ADMM scheme applied to the solution of the saddle-point problem (11), (12) is presented as follows:

$$u^k = \arg\min_{u \in \mathbb{R}^n} L(u, v^{k-1}, \xi^k), \tag{13}$$

$$v^k = \arg\min_{v \in \mathbb{R}^{2n}} L(u^k, v, \xi^k), \tag{14}$$

$$\xi^{k+1} = \xi^k - \tau(v^k - \nabla u^k). \tag{15}$$

In the following, we show briefly how to solve the two minimization sub-problems (13) and (14) for the primal variables u and v, respectively. Firstly, we present the resolution of the sub-problem (13).

Knowing the iterations v^{k-1} and ξ^k, the minimization sub-problem associated to the computation of u is given as follows:

$$u^k = \arg\min_{u \in \mathbb{R}^n} \left\{ \lambda ||u - f||_2^2 + < \xi^k, \nabla u > + \frac{\tau}{2} \|v^{k-1} - \nabla u\|_2^2 \right\}. \tag{16}$$

The resolution of this problem is in fact the iteration of a quadratic minimization problem where the first-order optimality conditions lead to the following linear system:

$$\left(I_n + \frac{\tau}{2\lambda} \Delta \right) u^k = f + \frac{\tau}{2\lambda} \operatorname{div} \left(v^{k-1} - \frac{1}{\tau} \xi^k \right), \tag{17}$$

where $\Delta \in \mathbb{R}^{n \times n}$ is the Laplacien matrix, which represents a 5-point stencil finite difference discretization of the negative 2D Laplace operator [13, 14]. div $\in \mathbb{R}^{n \times n}$ is the adjoint operator of the gradient ∇ such as: $\Delta = \operatorname{div}(\nabla)$. It follows that at any ADMM iteration the minimization sub-problem associated to the computation of u is in fact a resolution of a linear system that can be solved explicitly by one forward and one inverse 2D FFT transform.

For the resolution of the sub-problem for v, knowing u^k, ξ^k and using the formulation of the augmented Lagrangian functional in (11) the minimization sub-problem associated to v can be reformulated in the following form:

$$v^k = \arg\min_{v \in \mathbb{R}^{2n}} \left\{ \Psi(|v|) + \frac{\tau}{2} \|v - w^k\|_2^2 \right\}, \tag{18}$$

where the vector $w^k \in \mathbb{R}^{2n}$ is constant with respect to the variable v, it is defined by

$$w^k = \nabla u^k + \frac{1}{\tau}\xi^k, \tag{19}$$

with ∇ is the discrete version of the gradient and $\xi^k \in \mathbb{R}^{2n}$ is the Lagrange multipliers at iteration k. The minimization problem in (18) is nonconvex since Ψ is nonconvex. To handle this nonconvexity, we use the multi-stage convex relaxation technique described in [27]. First of all, we define a convex function H and a concave function $\bar{R}$ satisfying:

$$\Psi(|v|) = R(v) = \bar{R}(H(v)). \tag{20}$$

where

$$R(v) = \frac{|v|}{1+\rho|v|}, \quad \text{with} \quad H(v) = |v|^p \quad \text{and} \quad \bar{R}(z) = \frac{z^{\frac{1}{p}}}{1+\rho z^{\frac{1}{p}}}, \quad \text{where } z \in [0, \infty[.$$

To apply the multi-stage convex relaxation method, we have to compute $R^*(\widehat{v})$ which is the concave dual of $\bar{R}(z)$ such as:

$$R^*(\widehat{v}) = \min_z \left[- < z, \widehat{v} > + \bar{R}(z) \right]. \tag{21}$$

It is in fact easy to compute $\widehat{v}$ based on the duality proprieties of the function $\bar{R}(z)$ such as:

$$\widehat{v} = \frac{1/p|v|^{1-p}}{(1+\rho|v|)^2}. \tag{22}$$

By choosing $p = 1$ to avoid a zero denominator in $\widehat{v}$. Then, the dual function is defined by:

$$R^*(\widehat{v}) = \frac{1}{\rho}(\sqrt{\widehat{v}} - 1)^2, \tag{23}$$

with $\widehat{v} > 0$. Which give the explicit solution (see [25] for more details):

$$v^{k+1} = shrink(\widehat{v}^k, w^k); \tag{24}$$

where $shrink(x, y) = \dfrac{x}{\|x\|}\max(\|x\| - y, 0)$; We are now ready to summarize previous results in Algorithm 1 where we report the main steps of the proposed iterative ADMM scheme used for the segmentation process.

$$Err = \frac{\|u^{k+1} - u^k\|_{L^2(\Omega)}}{\|u^k\|_{L^2(\Omega)}}. \tag{25}$$

Data: Given an initial image f and the initial λ^0. Choose $tol, \rho, \tau > 0, \xi^0 > 0$
Result: The segmented image u and λ
begin
Compute λ `// Use (6)`
For $k = 1, 2, 3, \dots$ until convergence do:
update primal variables:
Compute u^k `// Use (17)`
Compute v^k `// Use (24)`
update dual variable:
Compute ξ^k `// Use (14)`
Set $k = k + 1$
Iterate until $Err < tol$.
`// Use (25)`

Algorithm 1: ADMM algorithm for image segmentation

4 Numerical Results

In this section we report some numerical experiments of the proposed algorithm tested through several grayscale images. The proposed method is first evaluated on shape matching/retrieval which is a very critical problem in computer vision. In particular, we demonstrate the ability of the model to generate large deformations and to produce a physically sound average shape. The algorithm is implemented using MATLAB and the computations have been made on an Intel Core i7 computer with 2.60 GHz and 8 GB memory. The values of used parameters are $\lambda^0 = 0.1$, $\rho = 0.5$, $\tau = 1.5$, $tol = 10^{-6}$ and $\xi = (1, \dots, 1)' \in \mathbb{R}^{2n}$.

The numerical experiments is based on a set of binary images from the MPEG7 shape database (http://www.dabi.temple.edu/shape/MPEG7/dataset.html). For each shape we have selected 20 images. For suck of simplicity we present 4 images for shape (device,cup,camel). As it can be inspected in Fig. 4, contrary to the existence approaches the non-convex regularization with non-smooth case, tends to better preserve the original contrast of the images and protects edges from over-smoothing.

In the Table 1 we present statistical validation metric of segmentation. Namely the Dice similarity coefficient (DSC) which evaluate the performance of both the reproducibility of manual segmentations and the spatial overlap accuracy of automated probabilistic fractional segmentation. The Jaccard index which is a statistic metric used for gauging the similarity and diversity of sample sets. Pixel accuracy which measure the percent of pixels in the image that correctly classified.

5 Conclusion

In this paper, we have introduced a novel variational segmentation approach using a non-convex regularization framework and a robust parameter selection to keep-safe image structures. In addition a fast ADMM algorithm is also investigated which

Fig. 4 The segmentation result of different shape images

Table 1 Statistical validation metric for the previous three examples

Image	Accuracy	Dice	Jaccard
Device	0.9917	0.9994	0.9987
	0.9909	0.9998	0.9981
	0.9907	0.9995	0.9979
	0.9893	0.9988	0.9967
Cup	0.9862	0.9886	0.9877
	0.9860	0.9883	0.9872
	0.9857	0.9878	0.9869
	0.9850	0.9871	0.9861
Camel	0.9998	0.9999	0.9994
	0.9995	0.9998	0.9993
	0.9993	0.9997	0.9992
	0.9991	0.9996	0.9990

ensure the convergence of the proposed method. Numerical results confirm the robustness of the proposed method.

Conflict of interest

- The authors declare that they have no conflict of interest.

References

1. M. V. Afonso, J. M. Bioucas-Dias, and M. A. Figueiredo. Fast image recovery using variable splitting and constrained optimization. IEEE transactions on image processing, 19(9):2345–2356, 2010
2. K. A. Berdawood, A. Nachaoui, R. Saeed, M. Nachaoui, and F. Aboud. An alternating procedure with dynamic relaxation for cauchy problems governed by the modified helmholtz equation. Advanced Mathematical Models & Applications, 5(1), 131–139, 2020
3. L. Chen, Y. Li, and T. Zeng. Variational image restoration and segmentation with Rician noise. J. Sci. Comput., 78(3):1329–1352, 2019
4. W. Dong, X. Li, L. Zhang, G. Shi, Sparsity-based image denoising via dictionary learning and structural clustering. In: CVPR 2011, pp. 457–464. IEEE (2011)
5. P. Dvalishvili, A. Nachaoui, M. Nachaoui, T. Tadumadze, On the well-posedness of the cauchy problem for a class of differential equations with distributed delay and the continuous initial condition. volume 43, pp. 146–160. Inst. Math. Mech. Natl. Acad. Sci. Azerbaijan 9 B Vahabzdeh (2017)
6. E. Ghadimi, A. Teixeira, I. Shames, and M. Johansson. Optimal parameter selection for the alternating direction method of multipliers (admm): quadratic problems. IEEE Transactions on Automatic Control, 60(3), 644–658, 2014
7. M. Hakim, A. Ghazdali, A. Laghrib. A multi-frame super-resolution based on new variational data fidelity term. Appl. Math. Model. (2020)
8. M. Hong and Z.-Q. Luo. On the linear convergence of the alternating direction method of multipliers. Mathematical Programming, 162(1–2), 165–199, 2017
9. M. Howard, M. C. Hock, B. T. Meehan, and L. E. Dresselhaus-Cooper. A locally adapting technique for edge detection using image segmentation. SIAM J. Sci. Comput., 40(4):B1161–B1179, 2018
10. M. Jung and M. Kang. Efficient nonsmooth nonconvex optimization for image restoration and segmentation. Journal of Scientific Computing, 62(2), 336–370, 2015
11. S.-H. Kim, K.-J. An, S.-W. Jang, and G.-Y. Kim. Texture feature-based text region segmentation in social multimedia data. Multimedia Tools and Applications, 75(20), 12815–12829, 2016
12. A. Laghrib, A. Ben-Loghfyry, A. Hadri, and A. Hakim. A nonconvex fractional order variational model for multi-frame image super-resolution. Signal Processing: Image Communication, 67:1–11, 2018
13. A. Laghrib, A. Chakib, A. Hadri, and A. Hakim. A nonlinear fourth-order pde for multi-frame image super-resolution enhancement. Discrete & Continuous Dynamical Systems-B, 25(1):415, 2020
14. A. Laghrib, M. Ezzaki, M. El Rhabi, A. Hakim, P. Monasse, and S. Raghay. Simultaneous deconvolution and denoising using a second order variational approach applied to image super resolution. Computer Vision and Image Understanding, 168:50–63, 2018
15. F. Li, C. Shen, and C. Li. Multiphase soft segmentation with total variation and H^1 regularization. J. Math. Imaging Vision, 37(2), 98–111, 2010
16. S. Lu, S.V. Pereverzev. *Regularization theory for ill-posed problems*, volume 58 of *Inverse and Ill-posed Problems Series*. De Gruyter, Berlin, 2013. Selected topics
17. Z. Lu, X. Jiang, G. Huo, D. Ye, B. Wang, Z. Zheng, A fast T-spline fitting method based on efficient region segmentation. Comput. Appl. Math. **39**(2) (2020)
18. A. Mitiche, I.B. Ayed, *Variational and Level Set Methods in Image Segmentation*, volume 5. Springer Science & Business Media (2010)
19. A. Nachaoui, M. Nachaoui, Iterative methods for forward and inverse bioelelectric field problem, in *International Conference on Applied Mathematics, Modeling and Life Sciences, Icamls' 18* (2018)
20. A. Nachaoui, M. Nachaoui, A. Chakib, M.A. Hilal, Some novel numerical techniques for an inverse Cauchy problem. J. Comput. Appl. Math. **381**, 113030, 21 (2021)
21. M. Nachaoui. Parameter learning for combined first and second order total variation for image reconstruction. Advanced Mathematical Models & Applications, 5(1), 53–69, 2020

22. S. S. Ngambeki, X. Ding, and M. D. Nachipyangu. Real time face recognition using region-based segmentation algorithm. Int. J. Eng. Res. Technol, 4(4), 875–878, 2015
23. S. Pare, A.K. Bhandari, A. Kumar, G.K. Singh, S. Khare, Satellite image segmentation based on different objective functions using genetic algorithm: A comparative study. in *2015 IEEE International Conference on Digital Signal Processing (DSP)*, pp. 730–734 (2015)
24. C. Van Chung, J. De los Reyes, C. Schönlieb, Learning optimal spatially-dependent regularization parameters in total variation image denoising. Inverse Problems **33**(7), 074005 (2017)
25. Y. Wang, J. Yang, W. Yin, and Y. Zhang. A new alternating minimization algorithm for total variation image reconstruction. SIAM Journal on Imaging Sciences, 1(3), 248–272, 2008
26. Y. Yuan and C. He. Variational level set methods for image segmentation based on both L^2 and Sobolev gradients. Nonlinear Anal. Real World Appl., 13(2):959–966, 2012
27. T. Zhang, Analysis of multi-stage convex relaxation for sparse regularization. J. Mach. Learn. Res. **11**(3) (2010)
28. F. Zhao and X. Xie. An overview of interactive medical image segmentation. Annals of the BMVA, 2013(7), 1–22, 2013

A Lagrangian Mixed Finite Elements Method for Advection-Diffusion Equations

Youness Mezzan and Moulay Hicham Tber

Abstract We present a computational methodology for solving advection-diffusion flow problems. Our method is based on an operator-splitting technique where the advection part is treated in a Lagrangian framework. The parabolic partial differential equation corresponding to the diffusive part is discretized using mixed finite elements. Therefore, the unknown concentration and the diffusive flux can be approximated simultaneously. The overall approach leads to a numerical diffusion-free and mass conservative scheme. We illustrate the performance of the proposed methodology using a representative numerical example.

Keywords Convection diffusion equations · Splitting scheme · Particle in cell method · Mixed finite elements

1 Introduction

Models featuring advection-diffusion equations arise in many areas ranging from fluid dynamics, heat and mass transfer to plasma physics and finance. It is well known that standard schemes applied to these equations produce solutions with non-physical oscillations in convection-dominated regimes. During the last decades, a variety of numerical techniques have been introduced to obtain better approximations [2]. The most popular mesh-based techniques (for instance streamline upwind Petrov-Galerkin methods) allow for an efficient treatment of the constitutive terms, but they affect the physical aspect by introducing numerical diffusions. Particle-based methods on the other hand are well suited but for pure convection problems only. To

Y. Mezzan
Department of Mathematics, Cadi Ayyad University, FSSM, Av. Prince My Abdellah, B.P. 2390, Marrakech, Morocco

M. H. Tber (✉)
Department of Mathematics, Cadi Ayyad University, FSTG, Av. Abdelkarim Elkhattabi, B.P 549, Marrakech, Morocco
e-mail: hicham.tber@uca.ac.ma

A. Nachaoui et al. (eds.), *Mathematical Control and Numerical Applications*, Springer Proceedings in Mathematics & Statistics 372,
https://doi.org/10.1007/978-3-030-83442-5_8

profit from the different features of the two approaches we propose a Lagrangian—Eulerian operator splitting scheme where the convective part is approximated along moving Lagrangian particles and the diffusion is solved with a mixed finite elements method in an Eulerian mesh. Throughout the simulation, the particle-mesh interactions are achieved using an efficient PDE-constrained optimization projection technique [4]. The overall scheme approximate the concentration and its corresponding flux simultaneously in a diffusion-free and locally conservative manner. A numerical example is presented to assess the performance of the proposed approach.

2 Eulerian-Lagrangian Method

Studying advection-diffusion problems gives place to a mathematical formulation in a domain $\Omega \subset \mathbb{R}^d$ with $d = 1, 2, 3$. The boundary Γ of Ω is assumed to be piecewise smooth. The unit outward normal to Γ is denoted by $\boldsymbol{n}$. We assume that the unknown of the problem is the concentration $c\,(x,\ t)$ with $x \in \Omega$ and $t \in (0,\ T)$ for some $T > 0$. The governing equation can be written as

$$\begin{aligned} &\tfrac{\partial c}{\partial t} + \boldsymbol{v} \cdot \nabla c - \nabla \cdot (\boldsymbol{D} \nabla c) = f && \text{in } \Omega \times (0, T), \\ &c = g && \text{on } \Gamma \times (0, T)\ , \\ &c(\cdot,\ t_0) = c_0 && \text{in } \Omega, \end{aligned} \tag{1}$$

where $\boldsymbol{v}(x, t)$ is a solenoidal advective velocity, $\boldsymbol{D}(x)$ is the diffusivity tensor and $f(x, t)$ is the volumetric source/sink. $c_0(x)$ stands for the initial concentration. For ease of exposition we consider Dirichlet boundary conditions with a prescribed concentration $g(x, t)$. The diffusivity tensor is symmetric, and is assumed to be bounded and satisfying the usual uniform ellipticity condition in Ω.

The Eulerian-Lagrangian method is conceived as an operator splitting procedure. We solve subsequently an advection subproblem

$$\begin{aligned} &\tfrac{\partial \psi}{\partial t} + \boldsymbol{v} \cdot \nabla \psi = 0 && \text{in } \Omega \times I^m, \\ &\psi = g && \text{on } \Gamma^- \times I^m, \\ &\psi = c\,(\cdot,\ t^m) && \text{in } \Omega \end{aligned} \tag{2}$$

and a diffusion one

$$\begin{aligned} &\tfrac{\partial c}{\partial t} - \nabla \cdot (\boldsymbol{D} \nabla c) = f && \text{in } \Omega \times I^m, \\ &c = g && \text{on } \Gamma \times I^m, \\ &c\,(\cdot,\ t^m) = \psi\,(\cdot,\ t^m) && \text{in } \Omega. \end{aligned} \tag{3}$$

The time interval $(0,\ T)$ is divided into M subintervals $I^m = \left(t^{m-1},\ t^m\right)$ with $t^m = m\Delta t$ for $m = 1, 2, \ldots, M$. Notice that in the advection stage, boundary conditions can only be prescribed at inflow boundaries denoted by Γ^-. $\mathcal{P}_L$ and $\mathcal{P}_E$ are projection operators, which are introduced in order to couple the fields ψ and c with these fields will be defined on particles and a mesh, respectively. Therefore, a single time step consists of the following steps:

1. solve advection Eq. (2) with a Lagrangian scheme at the particle level;
2. transfer the intermediate solution ψ from the particles onto the Eulerian mesh using the operator $\mathcal{P}_E$;
3. solve Eq. (3) using mixed finite elements on an Eulerian grid;
4. update the Lagrangian concentration ψ from c using the operator $\mathcal{P}_L$.

2.1 Lagrangian Discretization of the Advection Problem

The Lagrangian configuration is described by a set of moving particles which carry the physical properties (here the concentration) of the medium. At a fixed time instant t we define $\Psi_t = (\mathcal{X}_t, \mathcal{C}_t)$ with

$$\mathcal{X}_t := \left\{ \boldsymbol{x}_p(t) \in \Omega \right\}_{p=1}^{N_p},$$

and

$$\mathcal{C}_t := \left\{ \psi_p(t) = \psi\left(\boldsymbol{x}_p(t),\, t\right) \in \mathbb{R} \right\}_p^{N_p}.$$

N_p is the number of particles. $\boldsymbol{x}_p$ and c_p denote respectively the spatial coordinates and the associated concentration of a particle p. The advection equation (2) corresponds to

$$\begin{aligned} \frac{d\psi_p(t)}{dt} &= 0 && \text{in } I^m, \\ \frac{d\boldsymbol{x}_p(t)}{dt} &= \boldsymbol{v}\left(\boldsymbol{x}_p(t), t^{m-1}\right) && \text{in } I^m. \end{aligned} \tag{4}$$

By integrating the particle trajectories equations using a three-stage third-order Runge– Kutta scheme the particle position is advanced from $\boldsymbol{x}_p(t^{m-1})$ to $\boldsymbol{x}_p(t^m)$. The Lagrangian concentration of each particle as a function of time is constant $\psi_p(t^{m-1}) = \psi_p(t^m)$.

2.2 Eulerian Discretization of the Diffusion Subproblem

Using a backward Euler time integration, the weak formulation of (3) in the mixed form consists in finding $c^m \in L^2(\Omega)$ and $\boldsymbol{\sigma}^m \in H(\text{div})$ such that

$$\begin{cases} \left(\boldsymbol{D}^{-1}\boldsymbol{\sigma}^m,\, \boldsymbol{\tau}\right) + \left(c^m,\, \nabla\cdot\boldsymbol{\tau}\right) = < g,\, \boldsymbol{\tau}\cdot\boldsymbol{n} >_\Gamma & \forall \boldsymbol{\tau} \in H(\text{div}), \\ \left(\nabla\cdot\boldsymbol{\sigma}^m,\, w\right) - \left(\frac{c^m}{\Delta t},\, w\right) = -\left(f^m + \frac{c^{m-1}}{\Delta t},\, w\right) & \forall w \in L^2(\Omega), \end{cases} \tag{5}$$

where $(\cdot,\ \cdot)$ and $\langle\cdot,\ \cdot\rangle_\Gamma$ indicates the inner product on $L^2(\Omega)$ and $L^2(\Gamma)$ respectively. $w \in L^2(\Omega)$ and $\boldsymbol{\tau} \in H(\text{div})$ are arbitrary test functions. The Hilbert space $H(\text{div})$ is defined by

$$H(\text{div}) = \left\{\boldsymbol{\tau} \in L^2(\Omega)^2,\ \nabla\cdot\boldsymbol{\tau} \in L^2(\Omega)\right\}.$$

This weak formulation allows us to approximate the diffusion subproblem by using mixed finite element methods. To this end, let $\mathcal{T}_h$ be a finite element partition of Ω into triangles. We consider a stable pair $(\mathcal{V}_h,\ \mathcal{Q}_h)$ of spaces on $\mathcal{T}_h$ such that $c_h^m \in \mathcal{Q}_h \subset L^2(\Omega)$ and $\boldsymbol{\sigma}_h^m \in \mathcal{V}_h \subset H(\text{div})$. In the numerical experiments Raviart-Thomas elements of lowest order RT_0 is used for $\mathcal{V}_h$ with a piecewise constant pressure discretization. Our mixed finite element problem consists in finding a couple $(c_h^m,\ \boldsymbol{\sigma}_h^m)$ in $\mathcal{Q}_h \times \mathcal{V}_h$ satisfying

$$\begin{cases} \left(\boldsymbol{D}^{-1}\boldsymbol{\sigma}_h^m,\ \boldsymbol{\tau}_h\right) - \left(\nabla\cdot\boldsymbol{\tau}_h,\ c_h^m\right) = -< g,\ \boldsymbol{\tau}_h\cdot n >_\Gamma & \forall\boldsymbol{\tau}_h \in \mathcal{V}_h, \\ \left(\nabla\cdot\boldsymbol{\sigma}_h^m,\ w_h\right) - \left(\frac{c_h^m}{\Delta t},\ w_h\right) = -\left(f^m + \frac{c_h^{m-1}}{\Delta t},\ w_h\right) & \forall w_h \in \mathcal{Q}_h. \end{cases} \tag{6}$$

Using a standard argument (see for instance [3, 5]), problems (5) and (6) possess unique solutions. Further, the matrix form of (6) correspond to a standard saddle point problem which can be solved using a Schur complement decomposition technique (see [1]).

2.3 Particle–Mesh Interaction

The projection operator $\mathcal{P}_E : \Psi_t \to \mathcal{Q}_h$ transferring the scattered particle data to piecewise continuous fields at the mesh is formulated in terms of a least-squares projections obtained from the following minimization problem

$$\min_{c_h\in\mathcal{Q}_h} J = \frac{1}{2}\sum_{p=1}^{N_p}\left(c_h\left(\boldsymbol{x}_p(t), t\right) - c_p(t)\right)^2$$

with c_h being constrained by the advection problem (2).

2.4 Mesh-Particle Projection

The mesh-particle projection $\mathcal{P}_L : \mathcal{Q}_h \to \Psi_t$ corresponds to a simple interpolation

$$c_p(t) = c_h\left(\boldsymbol{x}_p(t), t\right) \qquad \text{for}\, p = 1, \ldots, N_p.$$

3 Numerical Experiments

We use a 2D Gaussian hill problem to experimentally validate the presented Lagrangian mixed finite element scheme. The computational domain Ω is a disk of center at the origin and redius $r = 0.5$. We make a partition $\mathcal{T}_h$ consisting of 1570 triangular elements and 833 vertices; see Fig. 1. We take $\boldsymbol{v} = \pi(-y, x)$, $T = \pi$, $f = 0$, $\boldsymbol{D} = \kappa = 10^{-2}$and $\Delta t = 0.01$. Initial and boundary conditions are deduced from the exact solution given by

$$c_{exact}(x, t) = \frac{2\sigma^2}{2\sigma^2 + 4\kappa t} \exp\left(\frac{\|\bar{x}(x, t) - x_c\|^2}{2\sigma^2 + 4\kappa t}\right)$$

with $\sigma = 0.1$, $'\bar{x}(x, t) = (\cos(\pi t)x + \sin(\pi t)y, \; -\sin(\pi t)x + \cos(\pi t)y)$ and $x_c = (-0.15, \; 0)$. In the Lagrangian step particles are generated such that each cell contains at least 5 particles. The saddle point problem in the Eulerian step is solved using a direct solver. In Fig. 2 we show the concentration profiles and the corresponding diffusive fluxes at $t = \pi/2$ and $t = \pi$. A good agreement between the computed and analytical quantities is achieved with an L^2-errror of at most 7.3×10^{-3}.

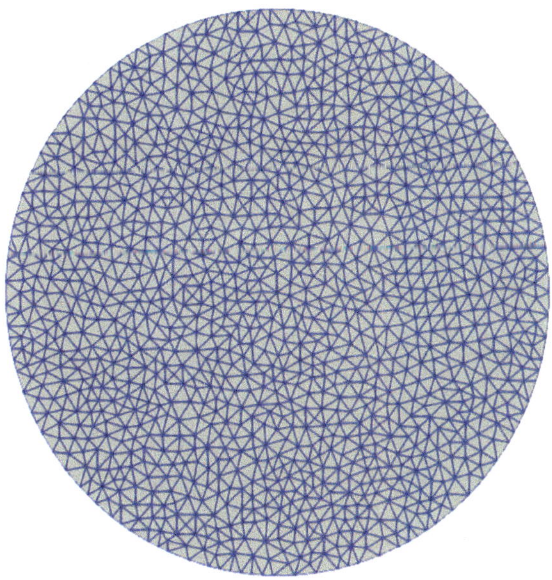

Fig. 1 Computational domain triangulation

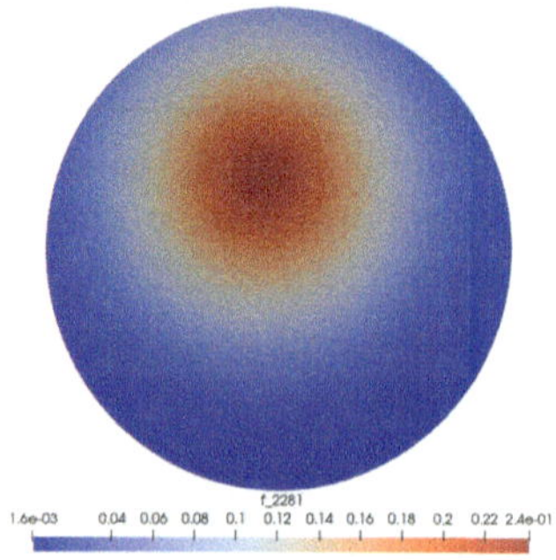

(a) exact concentraion at $t = \pi/2$

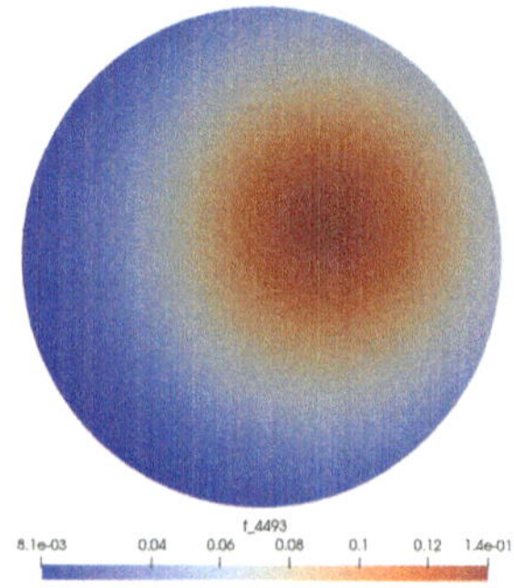

(b) exact concentration at $t = \pi$

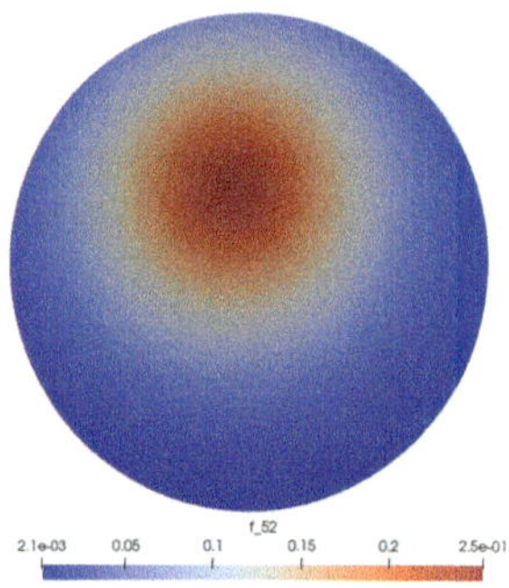

(c) computed concentraion at $t = \pi/2$

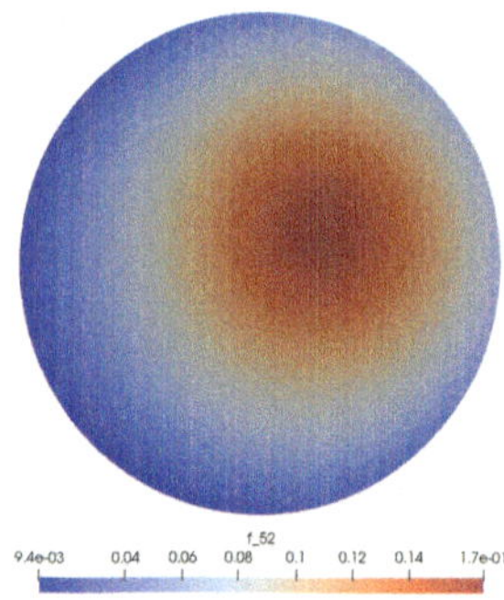

(d) computed concentraion at $t = \pi$

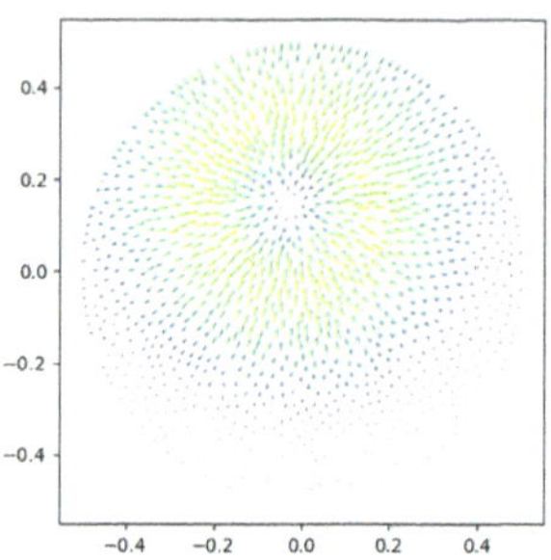

(e) computed diffusive flux at $t = \pi/2$

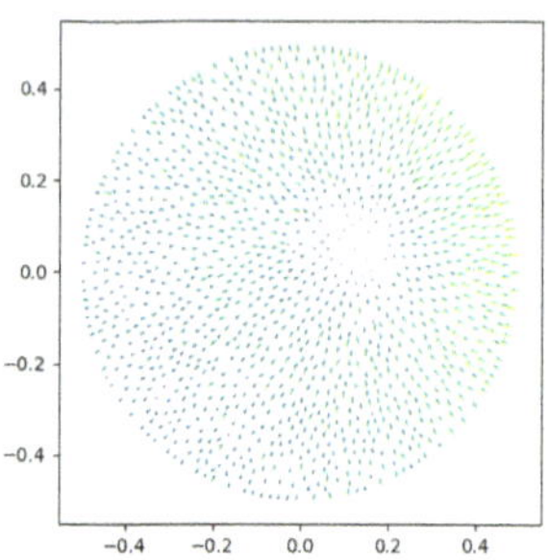

(f) computed diffusive flux at $t = \pi$

Fig. 2 2D Gaussian hill numerical experiment

References

1. M. Benzi, G.H. Golub, J. Liesen, Numerical solution of saddle point problems, Acta Numer. **14**, 1–137 (2005). https://doi.org/10.1017/S0962492904000212
2. J. Donea, A. Huerta, *Finite Element Methods for Flow Problems* (Wiley, 2003)
3. G.N. Gatica, *A Simple Introduction to the Mixed Finite Element Method: Theory and Applications, Springer Briefs in Mathematics* (Springer, Cham, Heidelberg, New York, Dordrecht London, 2014), p. 142
4. J.M. Maljaars, C.N. Richardson, N. Sime, LEoPart: a particle library for FEniCS. Comput. Math. Appl. **81**, 289–315 (2021). https://doi.org/10.1016/j.camwa.2020.04.023
5. P. Raviart, J.M. Thomas, A mixed finite element method for second order elliptic problems, in *The Mathmatics Aspects of Finite Element methods, Lecture Notes on Mathematics*, (Springer, Berlin), pp. 292–315. https://doi.org/10.1007/BFb0064470

Adaptation by Nash Games in Gradient-Based Multi-objective/Multi-disciplinary Optimization

J.-A. Désidéri

Abstract A two-phase numerical process is proposed for gradient-based multi-disciplinary optimization. In a first phase, one or several Pareto-optimal solutions associated with a subset of the cost functions, the primary objective functions, subject to constraints, are determined by some effective multi-objective optimizer. In the second phase, a continuum of Nash equilibria is constructed tangent to the Primary Pareto Front along which secondary cost functions are to be reduced, while best preserving the Pareto-optimality of the primary cost functions. The focus of the article is on estimating the rate at which the secondary cost functions are diminished. The method is illustrated by the numerical treatment of the optimal sizing problem of a sandwich panel w.r.t. structural resistance under bending loads and blast mitigation.

Keywords Gradient-based optimization · Multi-objective optimization · Multi-disciplinary optimization · Descent direction · Nash game · Adaptivity

AMS Classification 49M 49Q 90C 91A

1 Introduction

In engineering, the optimization of a complex system is almost invariably multi-objective, and this, for several reasons. Hence the problematics associated with multi-objective optimization has raised important developments in recent years [15].

Firstly, the problem is very often multi-disciplinary by the physics involved. To illustrate this, consider the prototypic example of the shape optimization of an aircraft wing, or configuration. Certainly the designer wishes to optimize the aerodynamic coefficients of the wing, six in total, since these impact directly the aircraft flight performance (mass, range, stability and maneuverability). However, during flight, the

J.-A. Désidéri (✉)
Université Côte d'Azur, 2004 Routes des Lucioles, B. P. 93, Centre Inria,
06902 Sophia Antipolis, Cedex, France
e-mail: jean-antoine.desideri@inria.fr

A. Nachaoui et al. (eds.), *Mathematical Control and Numerical Applications*,
Springer Proceedings in Mathematics & Statistics 372,
https://doi.org/10.1007/978-3-030-83442-5_9

configuration is submitted to important loads and it must be assessed also structurally, statically and dynamically. Additionally, an aircraft is not a rigid body and a proper assessment of the effects requires the coupling of these disciplines (aero-elasticity). Other considerations are also accounted for, and somewhat differently for civil or military aircraft in relation to acoustics, propulsion, stealth, etc.

The multi-disciplinary aspect has fostered important developments in Industry, since it impacts the way different groups within the enterprise cooperate to assemble a coordinated concept. Multi-level strategies have been proposed to address this question, and in particular the BLISS method [27] is widely used.

In relation to this, speciliazed computing architectures have also been devised [22].

Secondly, each discipline may require a multi-point treatment, since the system can be operated in several ways. In the aeronautical design, the entire flight envelope must be analyzed. In particular, for a civil aircraft, the optimization of take-off requires to maximize lift in the subsonic-flow regime (small Mach number, large angle of attack), whereas cruise requires the minimization of drag under lift constraint in transonic flow conditions (high Mach number, small angle of attack) for minimum kerosene consumption and air pollution. Hence numerical simulation serves to analysis and optimization in turns.

An additional major difficulty resides in the fact that the critical disciplines are most often evaluated via the integration of a system of partial-differential equations (PDE). For the aircraft aerodynamics alone, the steady Euler equations in three dimensions may be considered sufficient for drag minimization in transonic conditions, since the most important effect is related to the main shock wave. However, in the take-off phase, during which high-lift devices are deployed, the complexity of the flow requires a simulation of the complete Reynolds-Averaged Navier-Stokes (RANS) equations, in steady or time-dependent form. For structural analysis, simulations involve either a simple structural model, such as a beam model, [19], or a more general elasticity model embedded in a complete finite-element discretization (e.g. [18]; for acoustics, a wave equation using the pressure field as input; for stealth, the Maxwell equations; etc. Hence the cost functions, and the constraints, the latter sometimes causing even greater difficulty, are usually functionals and in large number.

Furthermore, in a complex engineering optimization, even when the set of potential important cost functions may be known a priori exhaustively by experience, in a multi-point optimization, after the treatment of a reduced subset of preponderant criteria, those additional objective functions that require a subsequent correction may depend on the operational point that is selected and be revealed by the process itself. Hence a certain form of adaptivity is required.

Concerning the PDE-constrained optimization aspect, differential elements (gradients and Hessians, mostly) can be calculated by adjoint approaches, e.g. [12], automatic differentiation, e.g. [16], or well-calibrated finite-differences. These issues are not discussed here. In the second phase of our algorithm, the question is simplified by the construction of either global or local quadratic meta-models.

Numerous contributions have been made in the field of evolutionary strategies, and more specifically using genetic algorithms. Perhaps one of the most employed is the Multi-objective Genetic Algorithm NSGA-II [3], but variants demonstrate specific merits [3, 17, 24, 25, 28]. These methods only require procedures for function evaluations, and are usually the most robust in finding global optima. This, of course, at the expense of a high computational demand. Nevertheless, such methods can indeed be applied in complex engineering problems [26]. In particular, in the framework of evolutionary algorithms, Greiner et al. [13] have documented a number of test-cases in structural engineering problems in which the multi-objective problematics is handled by a game formulation. They examined mostly Nash and Stackelberg games and their relationship with Pareto-optimality.

More recently, Bayesian Optimization (BO) methods have emerged as an alternative to evolutionary algorithms to solve global optimization problems using only function evaluations [20]. These statistical approaches are especially well suited to the case of expensive evaluations and are designed to account for limited computational budgets. In the context of multi-objective optimization, approximations of the Pareto front can be obtained with a relatively moderate number of evaluations [10], typically some dozens, whereas genetic algorithms require some hundreds of evaluations, and sometimes more.

While evolutionary strategies act on populations that are iteratively enriched by application of pseudo-Darwinian operators, classical methods operate on individual design points. Certain algorithms have been devised to directly enrich the Pareto front from a set of initial points, in particular:

- the normal boundary intersection [2] aiming to produce evenly-distributed points on the Pareto set, and related weights;
- the normalized normal constraint method [23], which incorporates an additional filter for a more proper identification;
- the Pareto-front interpolation [14], in which the authors construct a sub-complex of a Delaunay triangulation of a finite set of Pareto optimal outcomes, and devise special rules for checking the inherent non-dominance of complexes; the method, was further developed in various publications, e.g. [15], and is supported by a surrogate model to alleviate the high computational cost of function(al) evaluations.

Alternately, several authors have investigated the possibility of extending the steepest-descent method to multi-objective optimization. Among them, note the method of Fliege and Svaiter [11].

In this article, the choice of the effective multi-objective optimizer to conduct the first phase of optimization is left open. Our focus is here on the second phase of our proposed adaptation strategy, meant to reduce secondary cost functions, along a continuum of Nash equilibria, originating from a design-point of the Primary Pareto Front. In this way, a certain form of adaptation is introduced in the entire multi-objective optimization process.

Our adaptation strategy is defined step by step in the next section that also includes fundamental results related to convex hulls and the definition of a descent direction common to a set of cost functions, excluding the case of a weakly-Pareto optimal point. In Sect. 3, the theoretical analysis is developed and the existence of the Nash equilibria and the efficiency of the method are established. In particular, the negative directional derivatives of the secondary cost functions at the initial point on the continuum of Nash equilibria are evaluated. In Sect. 4, the method is illustrated by the numerical treatment of the optimal sizing of a structural element according to analytical models. As expected, the plot of cost functions normalized to their initial values exhibits a characteristic funnel-shaped pattern.

2 Nash-Game-Based Adaptation Algorithm

2.1 Mathematical Problem Definition

One considers a multi-objective differentiable optimization problem involving a set of M cost functions $\{f_j(\mathbf{x})\}$ ($j = 1, \dots, M$; $\mathbf{x} \in \Omega_{ad} \subseteq \mathbb{R}^n$), the first m ($1 \leq m < M$) of which are thought of preponderant, or primary, and the remaining ones ($j = m+1, \dots, M$), secondary, and all are subject to K equality constraints, $c_k(\mathbf{x}) = 0$ ($k = 1, \dots, K$). Objective and cost functions are all smooth, say $C^2(\Omega_{ad})$.

A first phase of optimization (omitted here) has led to the knowledge of a point $\mathbf{x}_A^\star$ in the open interior of the admissible domain Ω_{ad} that is Pareto-optimal for the minimization of the sole primary cost functions under the constraints.

The second phase of optimization is aimed to construct a continuum of Nash equilibria $\{\overline{\mathbf{x}}_\varepsilon\}$ parameterized by ε ($\varepsilon \geq 0$) in way such that: firstly, $\overline{\mathbf{x}}_0 = \mathbf{x}_A^\star$ (compatibility), and secondly, for ε sufficiently small, the Pareto-optimality condition of the primary cost functions is violated by a term $O(\varepsilon^2)$ only, whereas the secondary cost functions decrease at least linearly with ε.

2.2 Convex Hull, Common Descent Direction and Directional Derivatives

Problematics: Given m differentiable cost functions $\{f_j(\mathbf{x})\}$ ($j = 1, \dots, m$) ($\mathbf{x} \in \Omega_{ad} \subseteq \mathbb{R}^n$), and their gradients at some admissible point $\mathbf{x}_0 \in \Omega_{ad}$,

$$\mathbf{g}_j = \nabla f_j(x_0), \tag{1}$$

one seeks a vector $\mathbf{d}^\star \in \mathbb{R}^n$ in the direction of which the derivatives of all cost functions are positive:

$$D_{\mathbf{d}^\star} f_j(x_0) = (\mathbf{g}_j)^t \mathbf{d}^\star \geq 0 \quad (\forall j) \tag{2}$$

(superscript ^{t} for transposition). When $\mathbf{x}_0$ is not (weakly) Pareto-optimal (that is, Pareto-stationary), one such direction exists, and if a strict inequality holds in (2), the vector $(-\mathbf{d}^\star)$ provides a descent direction common to all cost functions.

To construct one such direction, we consider an $n \times n$ positive-definite matrix $\mathbf{A}_n$, and define the scalar product in $\mathbb{R}^n$

$$(\mathbf{u} \in \mathbb{R}^n) \quad (\mathbf{v} \in \mathbb{R}^n) \quad \big(\mathbf{u}, \mathbf{v}\big) = \mathbf{u}^t \mathbf{A}_n \mathbf{v} \tag{3}$$

and associated Euclidean norm

$$(\mathbf{u} \in \mathbb{R}^n) \quad \|\mathbf{u}\| = \sqrt{\big(\mathbf{u}, \mathbf{u}\big)} = \sqrt{\mathbf{u}^t \mathbf{A}_n \mathbf{u}}, \tag{4}$$

and consider the convex hull of the family of gradients:

$$\overline{\mathbf{U}} = \left\{ \mathbf{u} \in \mathbb{R}^n \text{ such that: } \mathbf{u} = \sum_{j=1}^{m} \alpha_j \mathbf{g}_j,\ \alpha_j \geq 0\ (\forall j),\ \sum_{j=1}^{m} \alpha_j = 1 \right\} \tag{5}$$

which evidently, is a closed, bounded and convex set.

Proposition 2.1 *The convex hull $\overline{\mathbf{U}}$ admits a unique element $\omega^\star$ of minimum Euclidean norm*

$$\omega^\star = \arg\min_{\mathbf{u} \in \overline{\mathbf{U}}} \|\mathbf{u}\|. \tag{6}$$

Proof Existence: the norm is continuous, and $\overline{\mathbf{U}}$ is closed and bounded. Uniqueness: $\overline{\mathbf{U}}$ is convex. □

Proposition 2.2 *Let $\mathbf{d}^\star = \mathbf{A}_n \omega^\star$. The elements $(\omega^\star, \mathbf{d}^\star)$ are such that:*

$$\forall \mathbf{u} \in \overline{\mathbf{U}} : \quad \big(\mathbf{u}, \omega^\star\big) = \mathbf{u}^t \mathbf{d}^\star \geq \|\omega^\star\|^2 \tag{7}$$

$$\forall j : \quad \big(\mathbf{g}_j, \omega^\star\big) = (\mathbf{g}_j)^t \mathbf{d}^\star \geq \|\omega^\star\|^2 \tag{8}$$

and if $\omega^\star \neq 0$, the vector $(-\mathbf{d}^\star)$ provides a common descent direction.

Proof Let $\mathbf{u} \in \overline{\mathbf{U}}$, arbitrary, and $\mathbf{v} = \mathbf{u} - \omega^\star$. Since $\overline{\mathbf{U}}$ is convex, one has:

$$\forall \theta \in [0, 1] : \ (1 - \theta)\omega^\star + \theta \mathbf{u} = \omega^\star + \theta \mathbf{v} \in \overline{\mathbf{U}}, \tag{9}$$

and since $\omega^\star$ is the minimum-norm element in $\overline{\mathbf{U}}$:

$$\|\omega^\star + \theta \mathbf{v}\|^2 = \big(\omega^\star + \theta \mathbf{v}, \omega^\star + \theta \mathbf{v}\big) \geq \|\omega^\star\|^2. \tag{10}$$

By developing the scalar product, one gets

$$\forall \theta \in [0,1]: \ 2\theta\big(\omega^\star, \mathbf{v}\big) + \theta^2 \|\mathbf{v}\|^2 \geq 0 \tag{11}$$

and this requires that the coefficient of θ, that is, $\big(\omega^\star, \mathbf{v}\big) = \big(\mathbf{u} - \omega^\star, \omega^\star\big)$, be positive, which yields the conclusion. □

Without great loss of generality, we now assume that the gradients are distinct. Let $\mathcal{F} = \{\mathbf{g}_j\}$ $(j = 1, \dots, m)$ denote their family. Let $\mathcal{F}_i$ be a subfamily of $\mathcal{F}$, $\overline{\mathbf{U}}_i$ its convex hull, and ω_i^* the minimum Euclidean norm element in $\overline{\mathbf{U}}_i$. Then, there is at least one subfamily $\mathcal{F}_i$ of minimal cardinal such that $\omega_i^* = \omega^\star$. Such a subfamily may be the sole $\mathcal{F}$ itself. Inversely, it may not be unique. Hereafter, one such subfamily will be said to be a "minimal generating subfamily" in the sense that it contains the minimal number of distinct elements permitting to generate $\omega^\star$ as a convex combination of its elements. Then, let $J^\star$ be the set of indices of the constitutive elements of the subfamily $\mathcal{F}_i$ so that:

$$\omega^\star = \sum_{j \in J^\star} \alpha_j^\star \mathbf{g}_j \tag{12}$$

in which $J^\star$ is a subset of $\{1, 2, \dots, m\}$, $\alpha_j^\star \geq 0$ $(\forall j \in J^\star)$ and $\sum_{j \in J^\star} \alpha_j^\star = 1$ for a convex combination. Furthermore, $\alpha_j^\star \neq 0$ $(\forall j \in J^\star)$ since the subfamily is of minimal cardinal; hence, $\alpha_j^\star > 0$ $(\forall j \in J^\star)$. Then:

Proposition 2.3 *Let $\sigma = \|\omega^\star\|^2$. The directional derivatives satisfy the following:*

$$D_{\mathbf{d}^\star} f_j = (\mathbf{g}_j)^t \mathbf{d}^\star = \sigma \ (\forall j \in J^\star), \quad \textit{and} \ \geq \sigma \ \textit{otherwise.} \tag{13}$$

Proof Let $\sigma_j = D_{\mathbf{d}^\star} f_j$ $(j \in J^\star)$, $\sigma_{\max} = \max_{(j \in J^\star)} \sigma_j$, and $\sigma_{\min} = \min_{(j \in J^\star)} \sigma_j$. Assume that $\sigma_{\max} \neq \sigma_{\min}$. Let $j_{\max} \in J^\star$ (resp. $j_{\min} \in J^\star$) be such that $\sigma_{j_{max}} = \sigma_{\max}$ (resp. $\sigma_{j_{min}} = \sigma_{\min}$). Recall that $\alpha_{j_{\max}}^\star > 0$, and consider strictly-positive numbers ϵ such that $0 < \epsilon < \alpha_{j_{\max}}^\star$, and let

$$\omega = \omega^\star - \epsilon \mathbf{g}_{j_{\max}} + \epsilon \mathbf{g}_{j_{\min}} = \sum_{j \in J^\star} \alpha_j \mathbf{g}_j \tag{14}$$

where $\alpha_j = \alpha_j^\star$ except for $\alpha_{j_{\max}} = \alpha_{j_{\max}}^\star - \epsilon > 0$ and $\alpha_{j_{\min}} = \alpha_{j_{\min}}^\star + \epsilon > 0$. Clearly, for sufficiently small and positive ϵ, $\alpha_j > 0$ $(\forall j)$, and $\sum_{j \in J^\star} \alpha_j = 1$. Hence $\omega \in \overline{\mathbf{U}}$. Additionally:

$$\begin{aligned} \|\omega\|^2 = \big(\omega, \omega\big) &= \left\|\omega^\star\right\|^2 - 2\epsilon\big(\omega^\star, \mathbf{g}_{j_{\max}} - \mathbf{g}_{j_{\min}}\big) + \epsilon^2 \left\|\mathbf{g}_{j_{\max}} - \mathbf{g}_{j_{\min}}\right\|^2 \\ &= \left\|\omega^\star\right\|^2 - 2\epsilon\left(\sigma_{\max} - \sigma_{\min}\right) + \epsilon^2 \left\|\mathbf{g}_{j_{\max}} - \mathbf{g}_{j_{\min}}\right\|^2 < \left\|\omega^\star\right\|^2 \end{aligned} \tag{15}$$

for all sufficiently small ϵ, and this is a contradiction since $\omega^\star$ is the minimum-norm element in $\overline{\mathbf{U}}$. The contradiction is removed by abandoning the hypothesis $\sigma_{\max} \neq \sigma_{\min}$. Hence $\sigma_j = \sigma = \|\omega^\star\|^2$ $(\forall j \in J^\star)$. Lastly, for $j \notin J^\star, \sigma_j \geq \sigma$ by virtue of Proposition 2.2. □

Proposition 2.4 *Let $\mathcal{F}_i$ be a minimal generating subfamily. Then, only two cases are possible:*

(1) either $\omega^\star = 0$ in a situation of Pareto-stationarity;
(2) or the subfamily $\mathcal{F}_i$ is linearly independent.

Proof By induction on m.

If $m = 2$, and if the two vectors are linearly dependent, either they are in opposite direction (Pareto stationarity), or the one of smaller norm constitutes alone the subfamily $\mathcal{F}_i$. Otherwise the two vectors form a linearly independent family.

Now, assume the statement is true when $m \leq \mu$, for some μ, and let us examine a case for which $m = \mu + 1$. In an affine-space representation, the vectors are given an origin O, and the convex hull associated with $\mathcal{F}_i$ is viewed as a polytope. If O belongs to the interior of this polytope, the gradients are in a situation of Pareto stationarity (1st case). Otherwise, $\omega^\star$ is identified by projection of O onto a boundary element, and the subfamily of vectors defining this boundary element constitutes a minimal generating subfamily and it contains at most $m - 1 \leq \mu$ elements: these, by inductive assumption, must be associated with linearly independent vectors. □

The configuration of the convex hull is illustrated in Appendix A in various cases of two and three gradients vectors.

Proposition 2.4 provides one of the possible ways to compute the element $\omega^\star$. In situations other than Pareto stationary, one way to proceed is to explore recursively the boundary of the polytope to identify a minimal generating subfamily. Then, one can evaluate the matrix $\mathbf{A}_n$ and the related metrics associated with the basis change from the canonical basis to this subset of gradients. In the new basis, $\omega^\star$ is associated with the bisector of the coordinate system and the back transformation yields $\omega^\star$ explicitly in the canonical basis [4, 5].

In all what follows, the standard Euclidean scalar product is used:

$$\mathbf{A}_n = \mathbf{I}_n, \quad \mathbf{d}^\star = \omega^\star. \tag{16}$$

2.3 Preliminary Calculations

The preliminary calculations are made of three main steps:

- Definition of a convex "Primary Steering Function" $f_A^+(\mathbf{x})$
- Definition of a "Territory Splitting"
- Definition of a "Secondary Steering Function" $f_B(\mathbf{x})$.

The cost functions are all assumed to be strictly positive. This may require the preliminary application of an adequate exponential transform, or the like. The exponential transform additionally permits us to consider gradients in the logarithmic form, and examine the variations of these functions in terms of proportions. Furthermore, this tends to improve convexity.

The Pareto-stationarity condition satisfied at $\mathbf{x} = \mathbf{x}_A^\star$ is expressed as follows:

$$\sum_{j=1}^{m} \alpha_j^\star \frac{\mathbf{P}\nabla f_j^\star}{f_j^\star} = 0 \tag{17}$$

where the supercript * on any symbol denotes throughout an evaluation at $\mathbf{x} = \mathbf{x}_A^\star$, ∇ is the gradient operator w.r.t. $\mathbf{x}$, $\{\alpha_j^\star\}$ $(j = 1, \ldots, m)$ are the known coefficients of a convex combination, $\mathbf{P}$ is the projection operator onto the linear subspace orthogonal to all constraint gradients, $\{\nabla c_k^\star\}$ $(k = 1, \ldots, K)$. Then the primary steering function is defined as follows:

$$f_A^+(\mathbf{x}) = \sum_{j=1}^{m} \alpha_j^\star \frac{f_j(\mathbf{x})}{f_j^\star} + \frac{c}{2} \left\| \mathbf{x} - \mathbf{x}_A^\star \right\|^2 \tag{18}$$

where c is a nonnegative constant adjusted to enforce the local convexity of $f_A^+(\mathbf{x})$ and the associated Lagrangian. Then, by virtue of (17), the primary steering function is stationary and convex at $\mathbf{x}_A^\star$, and therefore achieves a local minimum.

The territory splitting is based on the orthogonal decomposition of the following $n \times n$ reduced Hessian matrix [7, 8]:

$$\mathbf{H}'_A = \mathbf{P}\mathbf{H}_A^{+,\star}\mathbf{P} = \boldsymbol{\Omega}\mathcal{H}\boldsymbol{\Omega}^t \tag{19}$$

where $\mathbf{H}_A^{+,\star} = (\nabla\nabla^t f_A^+)^\star = \mathbf{H}_A^\star + c\mathbf{I}_n$ and $\mathbf{H}_A^\star = \nabla\nabla^t f_A^\star$, and $\nabla\nabla^t$ is the Hessian operator. Matrix $\mathbf{H}'_A$ is positive semi-definite, of rank $n - K$ exactly since matrix $\mathbf{H}_A^{+,\star}$ is strictly-positive definite by convexity fix. The eigenvectors, column-vectors of matrix $\boldsymbol{\Omega}$, are conventionally arranged in a special way: the first K are associated with the directions of the constraint gradients (null space of $\mathbf{P}$), and the remaining ones are ordered by decreasing eigenvalue. In this way, the tail eigenmodes are associated with thc smaller sensitivities of the primary steering function. The splitting is associated with the following change of variables:

$$\mathbf{x} = \mathbf{x}_A^\star + \boldsymbol{\Omega} \begin{pmatrix} \mathbf{u} \\ \mathbf{v} \end{pmatrix} := \mathbf{X}(\mathbf{u}, \mathbf{v}) \tag{20}$$

where $\mathbf{u} \in \mathbb{R}^{n-p}$, $\mathbf{v} \in \mathbb{R}^p$, and p is a free integer parameter to be chosen such that and $1 \le p < n - K$. In this setting, the following notations are used:

$$\boldsymbol{\Omega} = (\boldsymbol{\Omega}_\mathbf{u}\ \boldsymbol{\Omega}_\mathbf{v}), \quad \boldsymbol{\Omega}_\mathbf{u} = \begin{pmatrix} \boldsymbol{\Omega}_{\mathbf{uu}} \\ \boldsymbol{\Omega}_{\mathbf{vu}} \end{pmatrix}, \quad \boldsymbol{\Omega}_\mathbf{v} = \begin{pmatrix} \boldsymbol{\Omega}_{\mathbf{uv}} \\ \boldsymbol{\Omega}_{\mathbf{vv}} \end{pmatrix} \tag{21}$$

and the following scaled logarithmic gradients are associated with the secondary cost functions:

$$\mathbf{g}_j = \mathbf{S}^{-\frac{1}{2}} \frac{\nabla_{\mathbf{v}} f_j^\star}{f_j^\star} = \mathbf{S}^{-\frac{1}{2}} \frac{\boldsymbol{\Omega}_{\mathbf{v}}^t \nabla f_j^\star}{f_j^\star} \qquad (j = m+1, \ldots, M) \tag{22}$$

where the scaling matrix $\mathbf{S} = (\nabla\nabla_{\mathbf{vv}}^t f_A^+)^\star$ is the lower $p \times p$ diagonal block of matrix $\mathcal{H}$ of (19), positive-definite by convexity fix. Hence these gradients reflect the sensitivities of the secondary cost functions when $\mathbf{v}$ varies, and a scaling by elements related to the primary steering function as well. Following the strategy employed in the Multiple Gradient Descent Algorithm (MGDA) [7], one identifies the minimum-norm element ω_B^* in the convex hull of the gradients $\mathbf{g}_j$:

$$\omega_B^* = \sum_{j=m+1}^{M} \alpha_j^\star \mathbf{g}_j \tag{23}$$

and the coefficients $\{\alpha_j^\star\}$ of this convex combination are used to define the secondary steering function:

$$f_B(\mathbf{x}) = \sum_{j=m+1}^{M} \alpha_j^\star \frac{f_j(\mathbf{x})}{f_j^\star} \tag{24}$$

and this completes the preliminaries.

2.4 Nash Game Formulation and Statement of Theoretical Result

One defines a continuation parameter ε $(0 \le \varepsilon \le 1)$ and the auxiliary cost function:

$$f_{AB}(\mathbf{x}) = (1-\varepsilon) f_A^+(\mathbf{x}) + \varepsilon f_B(\mathbf{x}) \tag{25}$$

For every fixed value of ε, two virtual players, *Player A* and *Player B*, respectively controling the sub-vectors $\mathbf{u}$ and $\mathbf{v}$ in (20), are involved in a Nash game defined by the following strategies:

- Strategy A : *Player A* attempts to minimize $f_A^+\big(\mathbf{X}(\mathbf{u},\mathbf{v})\big)$ by the strategy $\mathbf{u}$, subject to the equality constraint $\mathbf{c}\big(\mathbf{X}(\mathbf{u},\mathbf{v})\big) = 0$, and by accounting for *Player B*'s fixed strategy $\mathbf{v}$,
  whereas:
- Strategy B: *Player B* attempts to minimize $f_{AB}\big(\mathbf{X}(\mathbf{u},\mathbf{v})\big)$ by the strategy $\mathbf{v}$, subject to no constraints, but by accounting for *Player A*'s fixed strategy $\mathbf{u}$.

The following results hold:

- The Nash equilibrium point $\overline{\mathbf{x}}_\varepsilon = \mathbf{X}\left(\overline{\mathbf{u}}_\varepsilon, \overline{\mathbf{v}}_\varepsilon\right)$ exists for all sufficiently small ε, and
$$\overline{\mathbf{x}}_0 = \mathbf{x}_A^\star \tag{26}$$
that is, $\overline{\mathbf{u}}_0 = 0 \in \mathbb{R}^{n-p}$ and $\overline{\mathbf{v}}_0 = 0 \in \mathbb{R}^p$, an essential property referred to as compatibility or consistency.
- Let $\phi_A(\varepsilon) = f_A^+\left(\overline{\mathbf{x}}_\varepsilon\right)$. Then $\phi_A'(0) = 0$ and
$$f_A^+ = 1 + O(\varepsilon^2) \tag{27}$$
which reflects the fact that, as ε varies, the Pareto-optimality of the primary cost functions is degraded by $O(\varepsilon^2)$ only.
- Let $\phi_B(\varepsilon) = f_B\left(\overline{\mathbf{x}}_\varepsilon\right)$. Then $\phi_B'(0) = -\sigma_B$, where $\sigma_B = \|\boldsymbol{\omega}^\star{}_B\|^2 \geq 0$. Hence, the secondary steering function $f_B(\mathbf{x})$ decreases linearly as ε increases, and by virtue of (13), all the secondary cost functions do as well, or faster.

Thus the objectives of our construction are achieved.

3 Proof and Convergence Rate

3.1 Compatibility

Proposition 3.1 *For $\varepsilon = 0$, the Nash equilibrium exists and corresponds to $\overline{\mathbf{x}}_0 = \mathbf{x}_A^\star$.*

Proof For $\varepsilon = 0$, $f_{AB} = f_A^+$ and both players attempt to reduce f_A^+, but by acting in different subdomains: *Player A* acts on the subvector $\mathbf{u}$, for fixed $\mathbf{v}$, and under the constraint $\mathbf{c}\big(\mathbf{X}(\mathbf{u}, \mathbf{v})\big) = 0$, whereas *Player B* acts on subvector $\mathbf{v}$, for fixed $\mathbf{u}$, and under no constraints. We have to prove that this concurrency results in the equilibrium
$$\mathbf{u} = \overline{\mathbf{u}}_0 = 0 \in \mathbb{R}^{n-p}, \quad \mathbf{v} = \overline{\mathbf{v}}_0 = 0 \in \mathbb{R}^p, \tag{28}$$
for which $\overline{\mathbf{x}} = \mathbf{x}_A^\star$. Equivalently: is $(\overline{\mathbf{u}}_0, \overline{\mathbf{v}}_0) = (0, 0) \in \mathbb{R}^{n-p} \times \mathbb{R}^p$ a solution to the following two uncoupled problems:
$$(A) \begin{cases} \min_{\mathbf{u}} f_A^+\big(\mathbf{X}(\mathbf{u}, 0)\big) \\ \text{s. t. } \mathbf{c}\big(\mathbf{X}(\mathbf{u}, 0)\big) = 0 \end{cases} \qquad (B) \begin{cases} \min_{\mathbf{v}} f_A^+\big(\mathbf{X}(0, \mathbf{v})\big) \\ \text{s. t. no constraints} \end{cases} \tag{29}$$

The affirmative answer is evident for *Subproblem (A)* which is a reformulation of the primary multi-objective problem in a subdomain where $\mathbf{u} = 0$ yields the solution $\mathbf{x}_A^\star$.

Concerning *Subproblem (B)*, note that it is the unconstrained minimization of a function that is convex by virtue of the convexity fix (18). Hence only stationarity remains to be verified for $\mathbf{v} = 0$, that is, $\mathbf{x} = \mathbf{x}_A^\star$. For this, consider an index $j \le p$:

$$\left(\frac{\partial f_A^+}{\partial v_j}\right)^\star = \nabla f_A^{+,\star} . \frac{\partial \mathbf{x}}{\partial v_j} = \nabla f_A^{+,\star} . [\boldsymbol{\Omega}]_{n-j+1} \tag{30}$$

where the notation $[\boldsymbol{\Omega}]_\ell$ is used for the ℓth column vector of matrix $\boldsymbol{\Omega}$, that is, the ℓth eigenvector of the reduced Hessian matrix $\mathbf{H}'_A$ of (19). Now, by optimality of $\mathbf{x}_A^\star$: $\nabla f_A^{+,\star} = -\sum_{k=1}^K \lambda_k \nabla c_k^\star \in Span\left\{\nabla c_k^\star\right\} = \mathrm{Ker}(\mathbf{P}) = Span\left\{[\boldsymbol{\Omega}]_1, \ldots, [\boldsymbol{\Omega}\}_K\right)$. Recall that the matrix $\boldsymbol{\Omega}$ is orthogonal, and its column vectors 2 by 2 orthogonal. Thus, provided that $K < n - p + 1$, $n - j + 1 > K$ and $[\boldsymbol{\Omega}]_{n-j+1}$ is orthogonal to this span and the above scalar product is indeed equal to 0. This confirms that *Subproblem (B)* does admit $\mathbf{v} = 0$ for trivial solution. □

3.2 Continuum of Nash Equilibria

Proposition 3.2 *Suppose that the following bounds are imposed on the subvectors* $\mathbf{u}$ *and* $\mathbf{v}$*:*

$$\|\mathbf{u}\| \le B_{\mathbf{u}}, \quad \|\mathbf{v}\| \le B_{\mathbf{v}}, \tag{31}$$

Then, the Nash equilibrium point exists for all sufficiently small ε.

Proof We begin by citing literally a theorem by Nash given in [1] (p. 268):

Theorem 3.3 (Nash) *Suppose that the multistrategy set*
(i) $X(N)$ *is a convex compact subset*
and that, for each player i*, the loss function*
(ii) f_i *is continuous and* $f_i(., \mathbf{x}^i)$ *is convex for all* $\mathbf{x}^i \in X^i$.
Then there exists a non-cooperative equilibrium.

Here, the strategy subsets of the two players are:

$$X^A = \left\{\mathbf{u} \in \mathbb{R}^{n-p} \;/\; \|\mathbf{u}\| \le B_{\mathbf{u}}\right\}, \quad X^B = \left\{\mathbf{v} \in \mathbb{R}^p \;/\; \|\mathbf{v}\| \le B_{\mathbf{v}}\right\}. \tag{32}$$

These are finite-dimensional closed balls, thus indeed convex compact subsets.

Player A, in effect, minimizes the Lagrangian associated to f_A^+ w.r.t. its strategy $\mathbf{u}$. By the adjustment of the convexity-fix constant c, this function is strictly convex, and so is also the corresponding partial function of this player's strategy.

Player B minimizes the cost function f_{AB}^+ w.r.t. its strategy $\mathbf{v}$. This function is strictly-convex for all sufficiently small ε, and so is also the corresponding partial function of this player's strategy.

Hence Theorem 3.3 applies, and this establishes Proposition 3.2. □

Furthermore, in all what follows, we assume that the dependency of the Nash equilibrium solution $\mathbf{x}_\varepsilon$ on the continuation parameter ε is smooth, and refer to the continuum of Nash equilibria as "the continuum":

Hypothesis 1 For strictly-positive but small enough ε, the Nash equilibrium solution $\overline{\mathbf{x}}_\varepsilon$, as a function of ε, is assumed to have C^2-regularity.

Remark 1 At the origin of the continuum, $(\bar{\mathbf{u}}_0, \bar{\mathbf{v}}_0) = 0$ and the bounds in (31) are satisfied trivially. Since continuity is assumed, this remains true for small enough ε. In this sense, these bounds are not really stringent.

3.3 *Variation of the Primary Cost Functions Along the Continuum*

Since the constraints

$$c_k\,(\overline{\mathbf{x}}_\varepsilon) = 0 \quad (\forall k) \tag{33}$$

are satisfied all along, by differentiating the above w.r.t. ε (denoted by $'$) and setting $\varepsilon = 0$ (denoted by the subscript $_0$), one gets the following initial directional derivatives:

$$(c_k)'_0 = \nabla c_k^\star . \overline{\mathbf{x}}'_0 = 0 \quad (\forall k). \tag{34}$$

But, due to the optimality condition,

$$\nabla f_A^\star = -\sum_{k=1}^{K} \lambda_k \nabla c_k^\star \quad (\lambda_k : \text{Lagrange multiplier}), \tag{35}$$

one gets accordingly:

$$(f_A)'_0 = \nabla f_A^\star . \overline{\mathbf{x}}'_0 = 0, \tag{36}$$

that is:

$$f_A = 1 + O(\varepsilon^2) \tag{37}$$

(and the same holds for f_A^+). Consequently, the Pareto-stationarity condition is degraded by a term $O(\varepsilon^2)$ only.

3.4 Variation of the Secondary Cost Functions Along the Continuum

3.4.1 Estimating $\overline{\mathbf{v}}$

By definition of *Subproblem (B)*:

$$\overline{\mathbf{v}} = \arg_{\mathbf{v}} \min f_{AB} \quad (\mathbf{u} \text{ fixed}) \tag{38}$$

in which f_{AB} is defined by (25) and expanded as follows in terms of $\mathbf{v}$ for fixed $\mathbf{u}$:

$$f_{AB}(\mathbf{v}) = f_{AB}(0) + \nabla_{\mathbf{v}} f_{AB}.\mathbf{v} + \frac{1}{2}\mathbf{v}.\nabla_{\mathbf{v}}\nabla_{\mathbf{v}}^t f_{AB}\mathbf{v} + O(\varepsilon^3). \tag{39}$$

Hence, optimality requires that

$$\nabla_{\mathbf{v}} f_{AB} + \nabla_{\mathbf{v}}\nabla_{\mathbf{v}}^t f_{AB}\mathbf{v} + O(\varepsilon^2) = 0 \tag{40}$$

which gives the following estimate:

$$\overline{\mathbf{v}} = -(\nabla_{\mathbf{v}}\nabla_{\mathbf{v}}^t f_{AB})^{-1}\nabla_{\mathbf{v}} f_{AB} + O(\varepsilon^2). \tag{41}$$

Now, as $\varepsilon \to 0$,

$$\nabla_{\mathbf{v}}\nabla_{\mathbf{v}}^t f_{AB} = (1-\varepsilon)\nabla_{\mathbf{v}}\nabla_{\mathbf{v}}^t f_A^+ + \varepsilon\nabla_{\mathbf{v}}\nabla_{\mathbf{v}}^t f_B = \mathbf{S} + O(\varepsilon) \tag{42}$$

where $\mathbf{S} = \left(\nabla_{\mathbf{v}}\nabla_{\mathbf{v}} f_A^+\right)^\star$. On the other hand:

$$\nabla_{\mathbf{v}} f_{AB} = (1-\varepsilon)\nabla_{\mathbf{v}} f_A^+ + \varepsilon\nabla_{\mathbf{v}} f_B \tag{43}$$

and: firstly, $\nabla_{\mathbf{v}} f_A^+ = O(\varepsilon^2)$ since f_A^+, as a function of ε, differs from 1 by $O(\varepsilon^2)$ uniformly; and secondly, $\nabla_{\mathbf{v}} f_B = \nabla_{\mathbf{v}} f_B^\star + O(\varepsilon)$. This gives:

$$\nabla_{\mathbf{v}} f_{AB} = \varepsilon\nabla_{\mathbf{v}} f_B^\star + O(\varepsilon^2). \tag{44}$$

Finally:

$$\overline{\mathbf{v}} = -\varepsilon\mathbf{S}^{-1}\nabla_{\mathbf{v}} f_B^\star + O(\varepsilon^2). \tag{45}$$

We now turn to the evaluation of matrix $\mathbf{S}$. Observe that:

$$\nabla_{\mathbf{v}} = \mathbf{\Omega}_{\mathbf{v}}^t\nabla, \quad \nabla_{\mathbf{v}}\nabla_{\mathbf{v}}^t = \mathbf{\Omega}_{\mathbf{v}}^t\nabla\nabla^t\mathbf{\Omega}_{\mathbf{v}} \tag{46}$$

$$\mathbf{S} = \mathbf{\Omega}_{\mathbf{v}}^t\mathbf{H}_A^{+\star}\mathbf{\Omega}_{\mathbf{v}} = \mathbf{\Omega}_{\mathbf{v}}^t\Big[\mathbf{P} + (\mathbf{I}_n - \mathbf{P})\Big]\mathbf{H}_A^{+\star}\Big[\mathbf{P} + (\mathbf{I}_n - \mathbf{P})\Big]\mathbf{\Omega}_{\mathbf{v}}. \tag{47}$$

Now, since $(\mathbf{I}_n - \mathbf{P})$ is the matrix associated with the orthogonal projection onto the subspace spanned by the constraint gradients, the column-vectors of matrix $\boldsymbol{\Omega}_{\mathbf{v}}$ are orthogonal to them. Consequently:

$$(\mathbf{I}_n - \mathbf{P})\boldsymbol{\Omega}_{\mathbf{v}} = 0_{n\times(n-p)}, \quad \boldsymbol{\Omega}_{\mathbf{v}}^t(\mathbf{I}_n - \mathbf{P}) = 0_{(n-p)\times n}, \tag{48}$$

so that

$$\mathbf{S} = \boldsymbol{\Omega}_{\mathbf{v}}^t \mathbf{P}\mathbf{H}_A^{+\star}\mathbf{P}\boldsymbol{\Omega}_{\mathbf{v}} = \boldsymbol{\Omega}_{\mathbf{v}}^t \mathbf{H}_A^{+'}\boldsymbol{\Omega}_{\mathbf{v}} = \boldsymbol{\Omega}_{\mathbf{v}}^t \boldsymbol{\Omega}\mathcal{H}\boldsymbol{\Omega}^t\boldsymbol{\Omega}_{\mathbf{v}} = \mathcal{H}_{\mathbf{v}} \tag{49}$$

in which $\mathcal{H}_{\mathbf{v}}$ is the "$\mathbf{v}$ diagonal block", that is, the lower $p \times p$ diagonal block of the positive-definite diagonal matrix $\mathcal{H}$ of (19). Hence the matrix $\mathbf{S}$ is diagonal positive-definite since $f_A^{+,\star}$ is convex.

3.4.2 Estimating $\overline{\mathbf{u}}$

By construction, the column vectors of matrix $\boldsymbol{\Omega}$ are the orthogonal eigenvectors of matrix $\mathbf{H}_A'$ of (19). The vector $\boldsymbol{\Omega}_{\mathbf{u}}\mathbf{u}$ is in the span of the first $n-p$ $(n-p > K)$ which contains the null space of the projection matrix $\mathbf{P}$ onto the subspace tangent to the constraint manifold. Thus, this vector is orthogonal to the constraint manifold, and the vector $\boldsymbol{\Omega}_{\mathbf{v}}\mathbf{v}$ is orthogonal to it. The sum of these two vectors is a "chord" for the manifold since both points $\mathbf{x}_A^\star$ and $\overline{\mathbf{x}}_\varepsilon$ satisfy the constraints. Consequently, assuming regularity of the manifold:

$$\|\boldsymbol{\Omega}_{\mathbf{u}}\overline{\mathbf{u}}\| = O\left(\|\boldsymbol{\Omega}_{\mathbf{v}}\overline{\mathbf{v}}\|^2\right). \tag{50}$$

Remark 2 This result describes how the Nash game does preserve the optimum of the primary cost function to second-order in ε: vector $\mathbf{u}$ alters f_A^+ and f_A by a term $O(\varepsilon^2)$, whereas vector $\mathbf{v}$ has no effect on it (to second-order) since $\boldsymbol{\Omega}_{\mathbf{v}}\mathbf{v} \perp \nabla f_A^{+\star} = \nabla f_A^\star$.

3.5 *Secondary Cost Functions*

The variation of the secondary cost functions is examined by letting:

$$\phi_j(\varepsilon) = \frac{f_j(\overline{\mathbf{x}}_\varepsilon)}{f_j^\star} \quad (j = m+1, \ldots, M). \tag{51}$$

This function is expanded in terms of ε using the established asymptotics:

$$\phi_j(\varepsilon) - 1 = \frac{1}{f_j^\star}\nabla f_j^\star.(\boldsymbol{\Omega}_{\mathbf{u}}\mathbf{u} + \boldsymbol{\Omega}_{\mathbf{v}}\mathbf{v}) + O(\varepsilon^2) = O(\varepsilon^2) - \frac{\varepsilon}{f_j^\star}\nabla f_j^\star.\boldsymbol{\Omega}_{\mathbf{v}}\mathbf{S}^{-1}\nabla_{\mathbf{v}} f_B^\star \tag{52}$$

and this provides the derivative:

$$\phi'_j(0) = -\frac{1}{f_j^\star}\nabla f_j^\star.\mathbf{\Omega}_\mathbf{v}\mathbf{S}^{-1}\nabla_\mathbf{v} f_B^\star = -\frac{1}{f_j^\star}(\nabla_\mathbf{v} f_B^\star)^t\mathbf{S}^{-1}\nabla_\mathbf{v} f_j^\star \tag{53}$$

To symmetrize this formula, the following new vector variable is introduced

$$\mathbf{w} = \mathbf{S}^{\frac{1}{2}}\mathbf{v} \tag{54}$$

so that $\nabla_\mathbf{w} = \mathbf{S}^{-\frac{1}{2}}\nabla_\mathbf{v}$ (- recall that $\mathbf{S}$ is positive-definite -) and (53) is rewritten:

$$\phi'_j(0) = -(\nabla_\mathbf{w} f_B^\star)^t\nabla_\mathbf{w}(f_j/f_j^\star) == -\mathbf{g}_j^t\omega_B^* \tag{55}$$

where the definitions made in (22)–(23) have been used. Lastly, the directional derivative is given by (13):

$$\phi'_j(0) = -\sigma_B \quad (j \in J_B^\star), \quad \leq -\sigma_B \text{ otherwise} \tag{56}$$

where $\sigma_B = \left\|\omega_B^*\right\|^2 \geq 0$, and $J_B^\star$ is a set of indices of minimal cardinal permitting to define the convex hull of the secondary gradients (w.r.t. $\mathbf{w}$).

Equation (56) establishes that our strategy for reducing the secondary cost functions linearly with ε is effective for small enough ε in all cases except if the gradients of the secondary cost functions w.r.t. the variable $\mathbf{w}$ are in a configuration of Pareto stationarity for which $\omega_B^* = 0$.

In effect, the steering function f_B forces all secondary cost functions to diminish at the same rate, at least, and this completes the proof of the theoretical statement. □

Remark 3 The definition of the variable $\mathbf{w}$, (54), is informative of how the negotiation between the two virtual players does operate. Intuitively, one might have guessed that the secondary cost functions react according to their partial gradients w.r.t. $\mathbf{v}$ since this vector constitutes *Player B*'s strategy. However, it is not exactly so. In effect, these partial gradients are scaled by the inverse square root of the block $\mathbf{S}$, and this reflects how *Player B* accounts for *Player A*'s strategy.

4 Illustration of the Prioritized Multi-objective Optimization Approach

4.1 Numerical Implementation

The present method can be applied formally to simple cases defined analytically. In the more general case in which the formal derivation is not at hand, we propose to resort to models of substitution, and formulate the Nash game in terms of these meta-models. Precisely, quadratic models can be constructed once for all for the two steering functions based on local differential elements at $\mathbf{x} = \mathbf{x}_A^\star$ (typically approximated by central differencing), and function values over a database to account on second derivatives more globally. For constraints, quadratic meta-models can also be constructed, but it is recommended to upgrade these models, as the determination of the continuum of Nash equilibria proceeds, because constraints participate critically to the definition of Pareto-optimality, and this notion cannot be altered legitimately. Such techniques have been implemented over the software platform http://mgda.inria.fr according to procedures fully defined in [6].

4.2 *Application to the Sizing of a Sandwich Element for Optimal Mechanical Resistance*

The present technique of prioritized optimization has been applied to four different test-cases with respect to 4 parameters that size an aluminum sandwich panel in order to maximize its mechanical performance in terms of strength under bending loads and blast mitigation [9]. Two of these test-cases are briefly summarized here. The reader is directed to the report for further methodological details and the precise data used in the numerical experiments.

4.2.1 Geometry

The structural element under consideration is a three-layer sandwich aluminum panel depicted in Fig. 1. The thicknesses of the upper and lower layers and core are respectively $\mathbf{t}_u$, $\mathbf{t}_l$ and $\mathbf{t}_c$. The following dimensions are used throughout $b = 5$ cm, $\ell = 50$ cm.

The core is made of honeycomb foam. Such structural elements are widely used by industrial manufacturers for their exceptional mechanical properties. Illustrations can be found in many engineering sites. In the present case, the honeycomb is symmetrical and its dimensions are defined Fig. 1.

In summary, the thicknesses $\mathbf{t}_u$, $\mathbf{t}_l$ and $\mathbf{t}_c$, as well as the ratio $\mathbf{R} = t/h$ of thickness to length of the honeycomb cell walls (see Fig. 1) constitute the design optimization variables.

4.2.2 Constraints

Expressing all sizes in meters, the design variables are subject to the following interval constraints

$$\mathbf{t}_u,\ \mathbf{t}_l \in [0.03, 0.14], \quad \mathbf{t}_c \in [0.05, 0.19], \quad \mathbf{R} \in [0.01, 0.20], \tag{57}$$

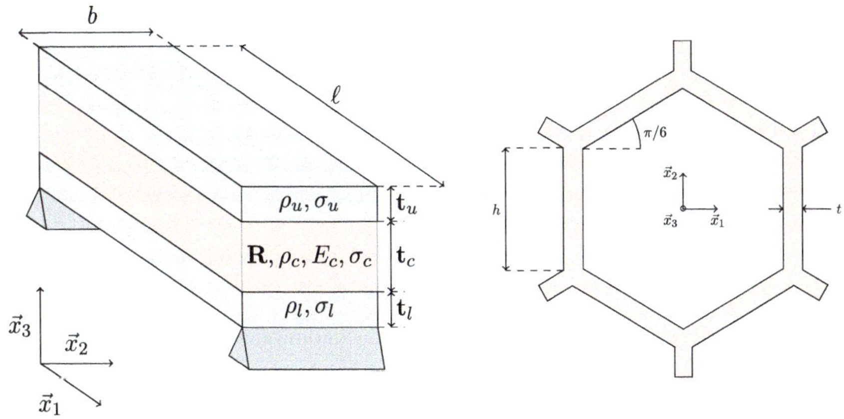

Fig. 1 Geometry of the sandwich panel (left) and honeycomb core (right)

and the total thickness to the following bound:

$$e = \mathbf{t}_u + \mathbf{t}_l + \mathbf{t}_c \leq 0.25. \tag{58}$$

4.2.3 Parameterization

In order to satisfy the interval constraints automatically, the following change of variables is adopted:

$$\begin{aligned} \mathbf{t}_u &= 0.085 + 0.055 \sin x_1 \\ \mathbf{t}_l &= 0.085 + 0.055 \sin x_2 \\ \mathbf{t}_c &= 0.120 + 0.070 \sin x_3 \\ \mathbf{R} &= 0.105 + 0.095 \sin x_4. \end{aligned}$$

where the variables x_i's are initially set in the interval $[-\pi/2, \pi/2]$ but may not remain in it in the course of the optimization.

Besides, in order to satisfy the bound constraint on the total thickness, an additional variable is introduced, x_5, as a slack variable, and the constraint is transformed into the following equality constraint:

$$c_1(\mathbf{x}) = 0.29 + 0.055\,(\sin x_1 + \sin x_2) + 0.070 \sin x_3 - 0.25 + x_5^2 = 0. \tag{59}$$

In this way, the problem involves 5 variables, and 1 constraint of equality type.

4.2.4 Cost Functions

The following five dimensionless cost functions have considered:

- φ_m representing mass (per unit transversal area);
- (φ_1, φ_2), inversely proportional to the critical failure forces $(F_{c,1}, F_{c,2})$ under bending loads, according to core indentation (1st mode) and exceeded lower-skin elastic resistance (second mode);
- φ_u, inversely proportional to the energy absorbed by the core under blast;
- φ_w, proportional to the panel central point deflection under blast.

All five cost functions are calculated from the geometrical definition, and thus in terms of $\mathbf{x}$, according to analytical models [9, 21].

4.2.5 Equating the Critical Flexural Forces

The first numerical experiment relates to mass and critical flexural forces only. Suppose that a first phase of optimization has consisted of the concurrent minimization of $f_1 = \varphi_m$ and $f_2 = \varphi_1$, under constraint (59), and resulted in the "Primary Pareto Front" represented by magenta symbols on Fig. 2. Then, the corresponding Pareto-optimal designs have been evaluated w.r.t. the secondary objective function $f_3 = \varphi_2$.

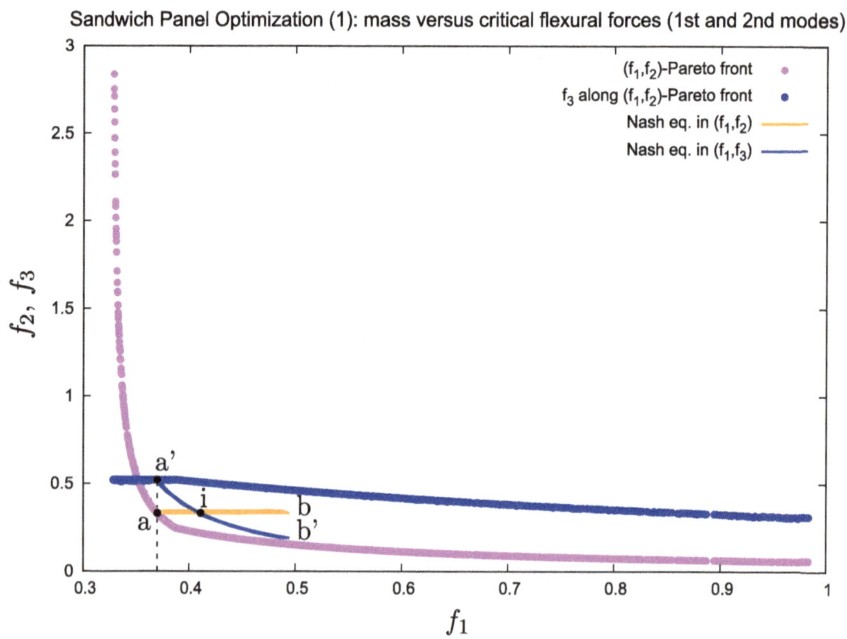

Fig. 2 Sandwich Panel Optimization (1): Primary Pareto Front (byzantine) and continuum of Nash equilibria (yellow) ($f_1 = \varphi_m,\ f_2 = \varphi_1,\ f_3 = \varphi_2$)

Table 1 Mass versus flexural forces - values of the objective functions at start and at the interpolated intersection point

Sandwich panel optimization (1)			
ε	f_1	f_2	f_3
0.	0.3702	0.3297	0.5194
0.6298	0.4111	0.3304	0.3304
	(+11.1%)	(+0.2%)	(−36.4%)

It then appears that for most of these designs, $f_3 \gg f_2$ and that is $F_{c,2} \ll F_{c,1}$. In other words, the second mode of failure can be triggered much prior to the first, and this not acceptable.

However, starting from such a design, the prioritized optimization procedure can be used to correct this deficiency, and achieve a design for which both critical forces, $F_{c,1}$ and $F_{c,2}$, are equal to a value only slightly inferior to the original value of $F_{c,1}$.

As an example, the (f_1, f_2)-Pareto-optimal design point for which $f_1 \approx 0.3702$ was considered, and defined as the starting point $\mathbf{x}_A^\star$ to initiate the computation of the continuum of Nash equilibria conducted to reduce f_3 using the MGDA platform [4]. The continuum of Nash equilibria is represented by the yellow path ab in terms of (f_1, f_2) and the blue path a'b' in terms of (f_1, f_3). These paths intersect at a point i where $f_1 \approx 0.4111$. There, mass has increased of some 11%, the first critical force has diminished marginally, of about 0.2%, the second critical force, inversely proportional to f_3, importantly increased of more than 57%, and, at point i, the two forces equate (see Table 1).

Remark 4 In a way, the prioritized approach has offered an alternative to the non-differentiable optimization of mass versus $\max_{\mathbf{x}} \min(F_{c,1}, F_{c,2})$.

4.2.6 Blast Mitigation

The second numerical experiment relates to blast mitigation. The "Primary Pareto Front" between mass, $f_1 = \varphi_m$, and the following blend of inverse critical flexural forces, $f_2 = \frac{1}{2}(\varphi_1 + \varphi_2)$, under constraint (59), has been determined first, and is represented by magenta symbols on Fig. 3. Then, the corresponding designs have been evaluated w.r.t. the following two secondary cost functions:

$$f_3 = \varphi_w, \quad f_4 = \frac{1}{2} e^{\varphi_u}. \tag{60}$$

(In f_4, the factor $\frac{1}{2}$ is for plotting convenience only; however the exponential transform is introduced to improve convexity [9], Appendix B).

Remark 5 Although the Pareto front has been determined densely and smoothly, the close examination of the Pareto set reveals that designs of different structures

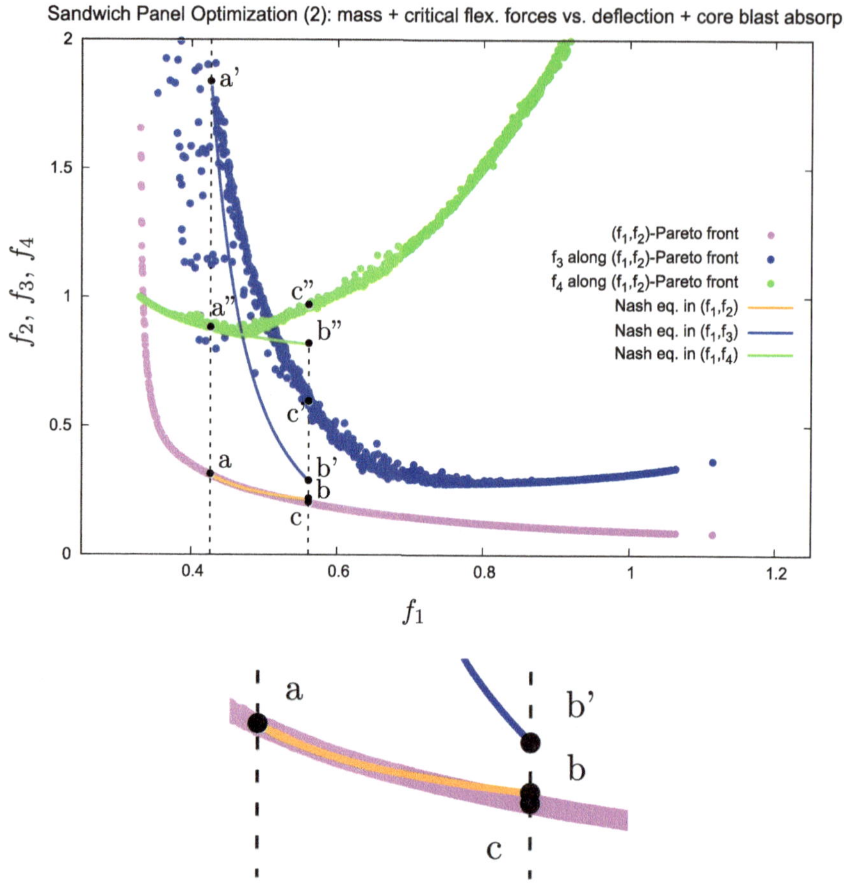

Fig. 3 Sandwich Panel Optimization (2): Primary Pareto Front (byzantine) and continuum of Nash equilibria (yellow) involving four objective functions ($f_1 = \varphi_m$, $f_2 = \frac{1}{2}(\varphi_1 + \varphi_2)$, $f_3 = \varphi_w$, $f_4 = \frac{1}{2}e^{\varphi_u}$ (top), and zoom of arc ab of continuum of Nash equilibria (below)

produce identical cost functions. As a result, the secondary cost functions "along the Pareto front" are multi-valued and exhibit an evident dispersion on the figure.

The second phase of optimization is launched from a design at which $f_1 \approx 0.4260$, whose performance is indicated on Fig. 3 by a, a' and a" in terms of f_2, f_3 and f_4 respectively. The continuum of Nash equilibria is represented by a yellow path ab in terms of (f_1, f_2), hardly distinguished from the Primary Pareto Front, a blue path a'b' in terms of (f_1, f_3), and a green path a"b" in terms of (f_1, f_4). At the endpoint, mass has increased of some 32%: the cost to pay. In return, the other three objective functions have been notably reduced: the average inverse critical flexural force of about 31% (point b), the deflection w of more than 84% (point b'), perhaps the most significant improvement, and f_4, based on the core blast energy absorption less,

Table 2 Blast mitigation versus Pareto-optimality between mass and the average of inverse critical flexural forces—values of the objective functions at start and near the end of the continuum ($\varphi_u^{-1} = 1/\ln(2f_4) \propto U_{\text{lost}}$)

Sandwich panel optimization (2)					
ε	f_1	f_2	f_3	f_4	φ_u^{-1}
0.	0.4260	0.3137	1.8380	0.8832	1.7576
0.9133	0.5602	0.2180	0.2885	0.8208	2.0175
	(+31.5%)	(−30.5%)	(−84.3%)	(−7.1%)	(+14.8%)

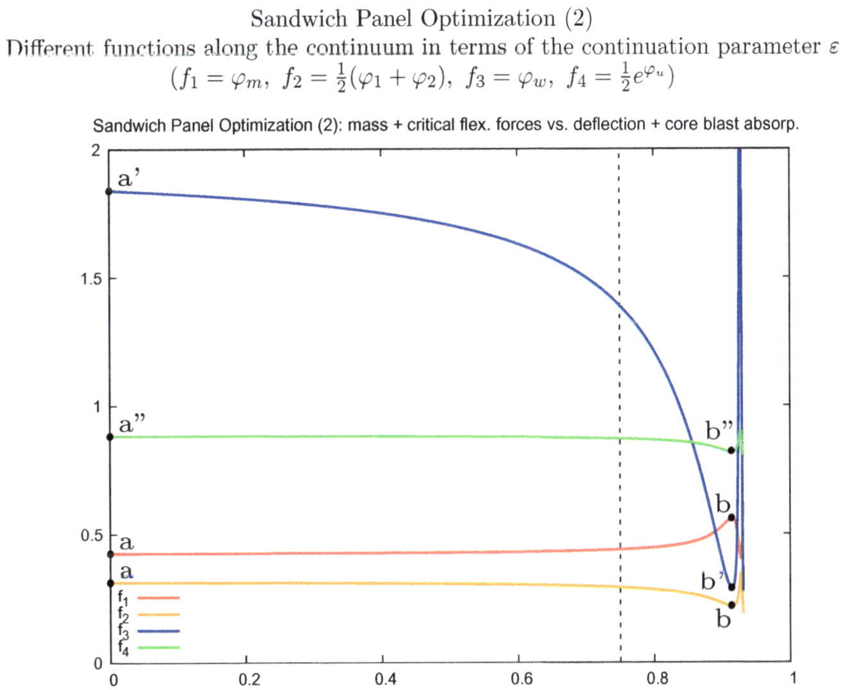

Fig. 4 Objective functions $\{f_j\}$

but significantly (7%) (point b"), corresponding to nearly 15% increase in energy absorption (see Table 2). Beyond this point, the variation of the cost functions in terms of the continuation parameter ε, indicated on Fig. 4, is no longer monotonic and eventually becomes unstable near the limit of convexity $\varepsilon_{\max}$ of the numerical formulation. These functions normalized to their respective initial values are represented on Fig. 5. This plot demonstrates the funnel-shaped pattern, a characteristic feature of the present approach:

- an increasing primary steering function with initial null slope, confirming that the Pareto-stationarity is degraded by a term $O(\varepsilon^2)$,

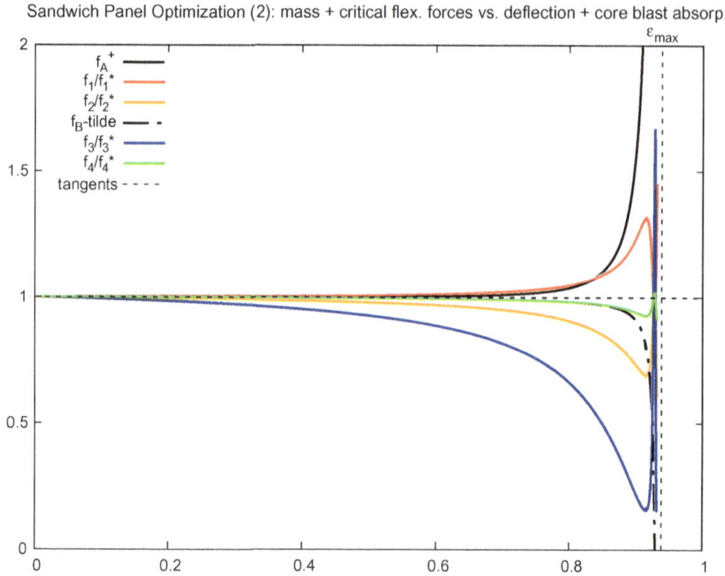

Fig. 5 Funnel-shaped pattern of objective functions as ratios to their initial values $\{f_j/f_j(\mathbf{x}_A^\star)\}$

- a decreasing secondary steering function with finite negative initial slope, $-\sigma_B$, and the identical behavior of all secondary cost functions.

The variation with ε of the optimization variables is given by Fig. 6. The Nash game affects all five variables. The core thickness increases and the ratio **R** here augments somewhat.

Figure 7 indicates the variation with ε of the constraint function $c_1(\mathbf{x})$. Apart from the very end of the continuum, when the instability is triggered, the constraint violation $\|c_1(\mathbf{x})\|$ remains less than 10^{-6}, two orders of magnitude inferior to the accuracy tolerance employed in this computation.

In conclusion, the prioritized optimization process was able to diminish both blast-mitigation-related objective functions. The reduction in deflection is impressive, the increase in energy absorption by the core more moderate, but significant. The numerical procedure demonstrates the general feature of the prioritized approach, in particular w.r.t. the funnel-shaped pattern of the cost functions.

5 Conclusion and Perspective

An adaptation technique based on Nash games has been proposed for multi-objective optimization. The technique operates after a first phase of multi-objective optimization has produced one, or several design points that are Pareto-optimal w.r.t. prepon-

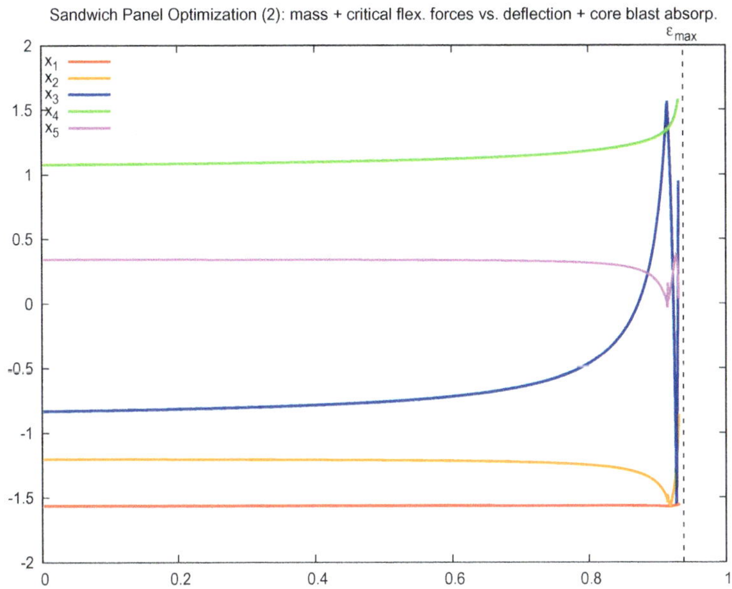

Fig. 6 Optimization variables $\{x_i\}$

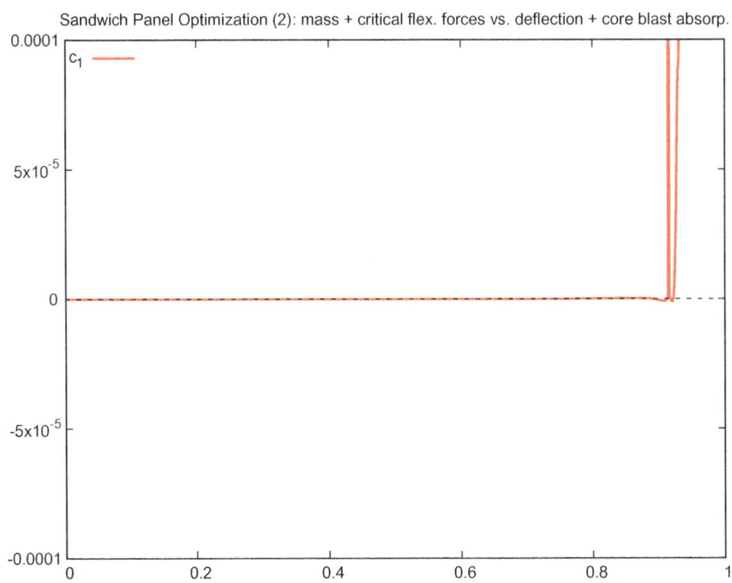

Fig. 7 Constraint function c_1

derant or primary cost functions under constraints. This prioritized approach permits to reduce additional or secondary cost functions in a second phase of optimization. A continuum of Nash equilibria, parameterized by ε, is constructed and its image in function space is tangent to the "Primary Pareto Front" which demonstrates that the Pareto-stationarity condition of the primary cost functions is degraded by $O(\varepsilon^2)$ only, while the secondary cost functions are reduced at least linearly with ε, with an initial negative directional derivative bounded from above by $-\sigma_B$. Thus the entire set of cost functions normalized to their values at start ($\varepsilon = 0$) exhibit a funnel-shaped pattern.

The theoretical development has permitted to relate the positive constant σ_B to gradients of the secondary cost functions w.r.t. a vector variable, $\mathbf{w}$, defined from an orthogonal decomposition of the domain referred to as "territory splitting", based on a reduced Hessian associated to the primary cost functions and the constraints. The definition of the variable $\mathbf{w}$ reflects in part the negotiation between the two virtual players defining the Nash game. The compatibility of the second phase of optimization to the result of the first phase has been justified, and the existence of the Nash equilibrium for small enough ε established.

To illustrate these theoretical results, a test-case in structural mechanics was solved numerically: a sandwich panel sizing problem meant to optimize the performance in terms of strength under bending loads, and blast mitigation, and involving four physical variables, one slack variable, one constraint, and up to four cost functions. The approach has permitted to identify Nash-equilibrium design points which, at the cost of a mass increase, were found, in terms of strength and blast mitigation, superior to the initial design point, while the trajectory of the continuum remains close to the Primary Pareto Front. The new design points were also found superior in efficiency to neighboring design points strictly on this Pareto front. This surprising result was alleged to the multi-valued nature of the secondary cost functions along the front.

Note that the same technique has also been successfully applied in [6] to a Supersonic Business Jet (SSBJ) sizing problem meant to optimize the aircraft flight performance in terms of take-off mass and distance, range, and approach speed involving 15 physical variables, one slack variable, and one constraint, yielding a similar behavior of the numerical method.

In the future, attempt will be made to apply the approach to a PDE-constrained optimization context in which the evaluation of the objective functions is necessarily more computationally demanding. In such a case, our plan is to develop additional local surrogate models based on statistical techniques and valid in the vicinity of the point chosen to initiate the continuum.

Acknowledgements The author wishes to express his warmest thanks to his colleagues from Inria and the University Côte d'Azur R. Duvigneau, A. Habbal, L. Hascoët, L. Monasse and B. Mourrain for very fruitful scientific discussions on Nash games and polytope exploration in large dimension, as well as software implementation of algorithms.

Special thanks are also due to the Inria Service of Experimentation and Development (SED), T. Kloczko, N. Niclausse and J.Wintz particularly, for the development of the web interface of the software platform http://mgda.inria.fr.

Appendix: Configuration of Convex Hulls

See the Fig. 8.

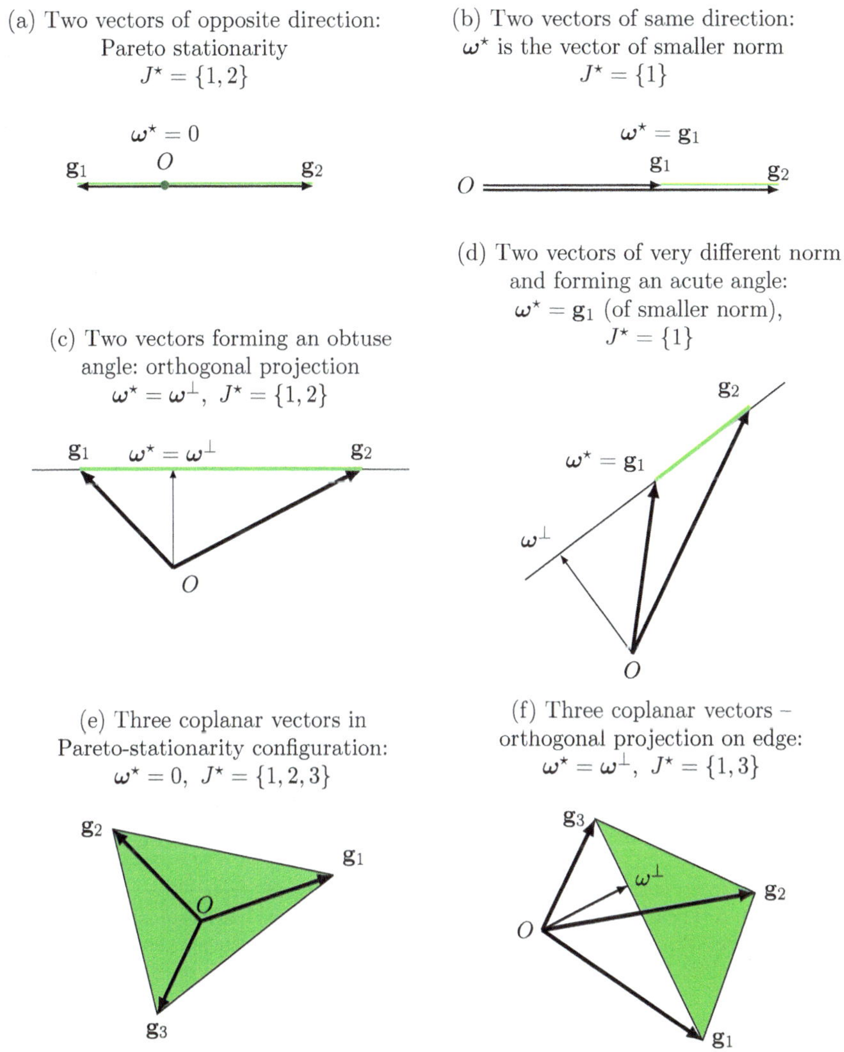

Fig. 8 Various geometrical configurations of the convex hull for two and three gradients vectors—In this affine-space representation, the vectors of the convex hull are given the origin O and are pointing over the green line or triangle

(g) Three coplanar vectors – non-orthogonal projection on vertex: $\boldsymbol{\omega}^\star = \mathbf{g}_2,\ J^\star = \{2\}$

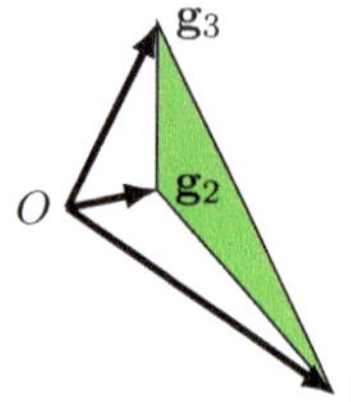

(h) Three l.i. vectors – orthogonal projection onto convex hull: $\boldsymbol{\omega}^\star = \boldsymbol{\omega}^\perp,\ J^\star = \{1, 2, 3\}$

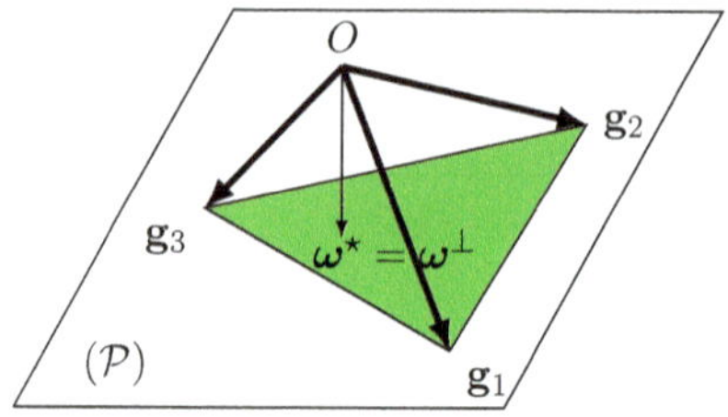

(i) Three l.i.vectors – orthogonal projection onto boundary edge: $J^\star = \{2, 3\}$

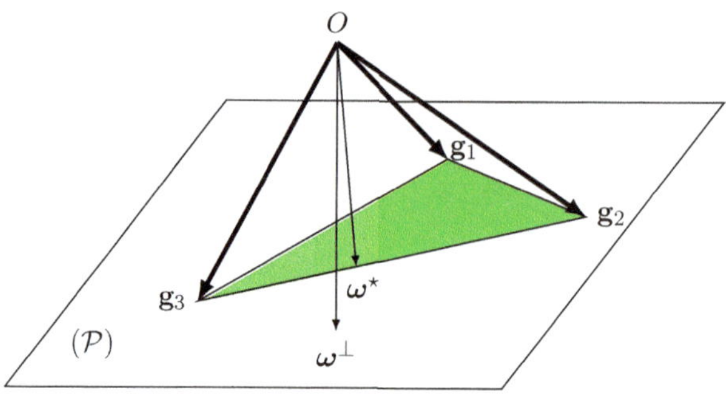

Fig. 8 (continued)

References

1. J.P. Aubin, *Mathematical Methods of Game and Economic Theory* (Courier Corporation, 2007)
2. I. Das, J. Dennis, *Normal-Boundary Intersection: An Alternate Method for Generating Pareto Optimal Points in Multicriteria Optimization Problems*, ICASE Report 96–62, ICASE (1996)
3. K. Deb, A. Pratap, S. Agarwal, and T. Meyarivan, A fast and elitist multiobjective genetic algorithm: NSGA-II, IEEE Transactions on Evolutionary Computation 6 (2002), pp. 182–197
4. J.A. Désidéri, *MGDA Software Platform for Multi-Objective Differentiable Optimization*, http://mgda.inria.fr
5. J.A. Désidéri, *Révision de l'algorithme de descente à gradients multiples (MGDA) par orthogonalisation hiérarchique*, Research Report 8710, Inria (2015). https://hal.inria.fr/hal-01139994
6. J.A. Désidéri, *Platform for Prioritized Multi-objective Optimization by metamodel-Assisted Nash Games*, Research Report 9290, Inria (2019). https://hal.inria.fr/hal-02285197
7. J.A. Désidéri, R. Duvigneau, *Direct and Adpative Approaches to Multi-objective Optimization*, Research Report 9291, Inria (2019). https://hal.inria.fr/hal-02285899
8. J.A. Désidéri, R. Duvigneau, DA. Habbal, *Computational Intelligence in Aerospace Sciences, V. M. Becerra and M. Vassile Eds.*, Progress in Astronautics and Aeronautics, T. C. Lieuwen Ed.-in-Chief Vol. 244, chap. Multi-Objective Design Optimization Using Nash Games, American Institute for Aeronautics and Astronautics Inc., Reston, Virginia (2014)

9. J.A. Désidéri, P. Leite, Q. Mercier, *Prioritized Multi-objective Optimization of a Sandwich Panel*, Research Report 9362, Inria, 2020. https://hal.inria.fr/hal-02931770
10. M. Emmerich, K. Yang, A. Deutz, H. Wang, C.M. Fonseca, *Advances in Stochastic and Deterministic Global Optimization, P. M. Pardalos and A. Zhigljavsky and J. Žilinskas eds.*, chap. A Multicriteria Generalization of Bayesian Global Optimization, Springer Optimization and Its Applications, Springer, Cham (2016)
11. J. Fliege, B.F. Svaiter, *Steepest Descent Methods for Multicriteria Optimization*, Math. Methods Oper. Rese., pp. 479–494 (2000)
12. M.B. Giles, N.A. Pierce, *An Introduction to the Adjoint Approach to Design, Flow, Turbulence and Combustion*, pp. 393–415 (2000)
13. D. Greiner, J. Périaux, J.M. Emperador, B. Galván, G. Winter, *Game Theory Based Evolutionary Algorithms: A Review with Nash Applicationss in Structural Engineering Optimization Problems*, Arch Comput Methods Eng (Springer), pp. 703–750 (2017)
14. M. Hartikainen and K. Miettinen, Constructing a pareto front approximation for decision making. Math. Methods Oper. Rese., 209–234 (2011)
15. M. Hartikainen, K. Miettinen, M.M. Wiecek, *PAINT: Pareto front interpolation for nonlinear multiobjective optimization*, Computational Optimization and Applications (2012), pp. 845–867
16. L. Hascoët, *Tapenade, version 3*, https://team.inria.fr/ecuador/fr/tapenade/
17. J. Horn, N. Nafpliotis, D.E. Goldberg, A niched Pareto genetic algorithm for multiobjective optimization, in *Proceedings of the First IEEE Conference on Evolutionary Computation. IEEE World Congress on Computational Intelligence* (1991)
18. A. Jameson, K. Leoviriyakit, and S. Shankaran, Multi-point Aero-Structural Optimization of Wings Including Planform Variations, in 45th Aerospace Sciences Meeting and Exhibit, Reno, Nevada, January 8–11. 2007
19. A. Jameson, J.C. Vassberg, S. Shankaran, Aerodynamic-structural design studies of low-sweep transonic wings. J. Aircraft, 47 (2010)
20. D.R. Jones, M. Schonlau, and W.J. Welch, Efficient global optimization of expensive black-box functions, Journal of Global Optimization 13 (1998), pp. 455–492
21. P. Leite, *Conception architecturale appliquée aux matériaux sandwichs pour propriétés multifonctionnelles*, Ph.D. diss., Université de Grenoble, École Doctorale I-MEP2, 2013. http://www.theses.fr/2013GRENI053
22. J.R.R.A. Martins, A.B. Lambe, Multidisciplinary design optimization: A survey of architectures. AIAA J., 51 (2013)
23. A. Messac, A. Isamail-Yahaya, C. Mattson, The normalized normal constraint method for generating the pareto frontier. Struct. Multidisciplinary Optim., 86–98 (2003)
24. T. Murata, H. Hishibuchi, MOGA: multi-objective genetic algorithms, in *Proceedings of 1995 IEEE International Conference on Evolutionary Computation, Perth, Australia*, pp. 289–294 (1995)
25. K. Parsopoulos, M. Vrahatis, Recent approaches to global optimization problems through particle swarm optimization. Natural Comput., 235–306 (2002)
26. J. Périaux, F. Gonzalez, D.S.C. Lee, Evolutionary optimization and game strategies for advanced multi-disciplinary design - applications to aeronautics and UAV design. Intelli. Syst. Control Autom. Sci. Eng. 75, Springer (2015)
27. J. Sobieszczanski-Sobieski, T.D. Altus, M. Phillips, R. Sandusky,Bilevel integrated system synthesis for concurrent and distributed processing. AIAA J., 41 (2003)
28. E. Zitzler and L. Thiele, Multiobjective evolutionary algorithms: A comparative case study and the strength Pareto approach, IEEE Transactions on Evolutionary Computation 3 (1999), pp. 257–271

Blind Noisy Mixture Separation for Dependent Sources

A. Ghazdali, A. Metrane, and A. Ourdou

Abstract The framework of this chapter is to introduce a new efficient Blind Source Separation (BSS) method that handles mixtures of noise contaminated independent/dependent sources. This approach is based on the minimization of a regularized criterion. Specifically, it consists in combining the total variation method to de-noise the observations, with the Kullbak-Leibler divergence between the copula densities to separate the mixtures. This algorithm has shown its effectiveness and efficiency toward the noisy dependent/independent sources and also surpassed the standard BSS algorithms through different experimental results.

Keywords Blind source separation · Dependent sources · Noisy sources · Copulas · Total variation regularization

1 Introduction

In a large number of situations or real world problems, we are dealing with signals made up of a mixture of many elementary primitive signals that would be useful to separate and get their initial form. To carry out this separation, an identification of the primitive signals must be made. However in most cases, only the mixed signals are observed, and we have little to no prior knowledge about neither the system of mixture nor the primitive or otherwise called the original signals.

Blind Source separation (BSS) [1] is the name given to the various and multiple statistical and deterministic methods used to restore the unobservable original signals called signal sources from a set of observed signals called observations (mixtures of the unknown sources) without any a priori knowledge, neither on the mixing system, nor on the sources. No information is therefore available, given these ambiguities, this problem is a highly ill-posed problem; it is difficult, if not impossible, to reconstruct the sources without making some assumptions. Most techniques depends on the assumption of the statistical independence of the sources such as Independent

A. Ghazdali (✉) · A. Metrane · A. Ourdou
LIPIM, ENSA-Khouribga, Sultan Moulay Slimane University, Beni Mellal, Morocco

A. Nachaoui et al. (eds.), *Mathematical Control and Numerical Applications*,
Springer Proceedings in Mathematics & Statistics 372,
https://doi.org/10.1007/978-3-030-83442-5_10

Component Analysis (ICA) [2–4], however the proposed approach handles even the dependent sources.

In this chapter, we will introduce a new blind source separation approach that handles the case when the mixture of the sources is noise-contaminated for both cases where the sources are statistically independent and dependent which make the resolution of the problem extra difficult. In the following we will present the BSS model then start detailing the new approach.

2 Problem Statement

Consider the problem of blind source separation in the case of a noisy instantaneous linear mixture. The noisy mixture is then written as

$$\bar{\boldsymbol{x}}(t) := \boldsymbol{\mathcal{A}}\boldsymbol{s}(t) + \boldsymbol{n}(t) \in \mathbb{R}^p, \tag{1}$$

where $\boldsymbol{s} \in \mathbb{R}^p$ is the vector of the sources, $\boldsymbol{n} \in \mathbb{R}^p$ is a measurement noise that is assumed to be Gaussian and independent of the sources, and $\boldsymbol{s} \in \mathbb{R}^{p\times p}$ is the unknown mixing matrix. The model (1) can also be written as

$$\bar{\boldsymbol{x}}(t) := \boldsymbol{x}(t) + \boldsymbol{n}(t), \tag{2}$$

where $\boldsymbol{x}(t) = \boldsymbol{\mathcal{A}}\,\boldsymbol{s}(t)$ is the vector of observations without noise.

The presence of noise in the mixture model complicates the source separation problem. The estimator of the sources $\boldsymbol{s}$ from the observations $\boldsymbol{x}$, in this case, is of the form

$$\bar{\boldsymbol{y}}(t) := \boldsymbol{\mathcal{B}}\bar{\boldsymbol{x}}(t), \tag{3}$$

where $\bar{\boldsymbol{y}}$ is the noisy estimate of the sources $\boldsymbol{s}$ and $\boldsymbol{\mathcal{B}}$ is the separation matrix. If we solve this problem by the classical method without taking into account the noise, the estimated sources would also be polluted by the noise. Indeed, the sources obtained according to the system (3) above, can be written as follows

$$\begin{aligned} \bar{\boldsymbol{y}}(t) &:= \boldsymbol{\mathcal{B}}\bar{\boldsymbol{x}}(t) \\ &= \boldsymbol{\mathcal{B}}\boldsymbol{x}(t) + \boldsymbol{\mathcal{B}}\boldsymbol{n}(t) \\ &= \boldsymbol{y}(t) + \bar{\boldsymbol{n}}(t). \end{aligned} \tag{4}$$

The estimated sources with noise are the sum of $\boldsymbol{y} = \boldsymbol{B}\boldsymbol{x}$ the estimate of the sources "without noise", and the noise $\bar{\boldsymbol{n}} = \boldsymbol{B}\boldsymbol{n}$. Ideally, we would like to estimate $\boldsymbol{y}(t)$ from (4) by denoising $\bar{\boldsymbol{y}}(t)$, but it is difficult, since the noise $\bar{\boldsymbol{n}}(t)$ is unknown. In the last few years, several algorithms have been proposed to solve the problem of blind source separation in the case of a noisy mixture [5–7]. The noise is likely to significantly modify the observations and thus lead to a bad estimation of the mixture matrix and the sources. In the next sections we will describe the copula based BSS approach for noisy mixtures [8]. This approach was based of a noise free copula based method [9] that later on was extended to the case of convolutive mixtures [10] and then in 2021 used the α-divergence instead of the kullback-leibler divergence to better improve the separation results [11]. These copula-based approaches were used for multiple real world applications such as the extraction of fetal ECG from the dependent abdominal signals [12] and digital image restoration [13].

3 First Step: Denoising of Observations

Generally, a signal is subject to a set of disturbances caused by several phenomena, leading to a degradation of its quality which makes its processing very difficult. Therefore, a pre-processing of the signals is highly necessary in most cases. Considering the requirements in several domains concerning the results obtained by denoising algorithms, the quality of such a pre-processing must be irreproachable: we must then eliminate the disturbing influences, while preserving the essential characteristics of the signals.

Given an observed signal $\bar{\boldsymbol{x}} = (\bar{x}_1, \ldots, \bar{x}_p)^\top \in \mathbb{R}^p$, and $\boldsymbol{x} = (x_1, \ldots, x_p)^\top \in \mathbb{R}^p$ the original signal. Let us suppose that $\boldsymbol{x}$ is degraded by an additive noise $\boldsymbol{n}$. The general denoising problem consists in being able to reconstruct $\boldsymbol{x}$ from the degraded signal $\boldsymbol{x}$. The relation between the observed signal $\boldsymbol{x}$ and the original signal $\boldsymbol{x}$ is given as follows:

$$\bar{\boldsymbol{x}}(t) = \boldsymbol{x}(t) + \boldsymbol{n}, \;\; \forall t \in \Omega. \tag{5}$$

with $\Omega = [0, T]$, where T is the observation time of the signals.

An intuitive way to find the signal $\boldsymbol{x}$ is to try to minimize the noise, which leads us to solve the following minimization problem:

$$\inf_{\boldsymbol{x}} \parallel \bar{\boldsymbol{x}} - \boldsymbol{x} \parallel^2_{L^2(\Omega)} . \tag{6}$$

The problem (6) is ill-posed, so it is not clear apriori that it has a solution on a well chosen space. Moreover and in a formal way, if we suppose that the problem (6) admits a solution $\tilde{\boldsymbol{x}}$, it must verify the following Euler Lagrange equation

$$\tilde{\boldsymbol{x}}(t) - \bar{\boldsymbol{x}}(t) = 0, \;\; \forall t \in \Omega. \tag{7}$$

From the equation (7), we notice that $\tilde{\boldsymbol{x}} + \epsilon$ is also a solution of the problem (6), which leads to a numerical instability

A classical way to remedy this kind of problem is to use regularization techniques. This will lead to a well-posed problem, where the existence and uniqueness of the solution is well assured. Among the most widely used are the Tikhonov type regularizations [14], and the Total Variation TV type regularizations [15, 16].

3.1 Regularization Total Variation

The great advantage of considering TV regularizations is to end up with convex functions with respect to the variable $\boldsymbol{x}$, this allows to use many tools of convex optimization to stop smoothing. The $\boldsymbol{x}$ signal restoration problem is now defined as follows

$$\inf_{x} \{ \frac{1}{2} \int_{\Omega} |\boldsymbol{x}(t) - \bar{\boldsymbol{x}}(t)|^2 \, dt + \lambda \int_{\Omega} \phi(|\nabla \boldsymbol{x}(t)|) \, dt \}. \tag{8}$$

The first term of (8) measures the distance of $\boldsymbol{x}$ with respect to the data, the second is the smoothing term which controls the variation of $\boldsymbol{x}$. The parameter λ is a positive constant which controls the smoothing. A strong noise requires a large value of regularization parameter λ. ϕ is a function of $\mathbb{R}^+$ with values in $\mathbb{R}^+$. It is well chosen to improve the smoothing in regions where the signal variations are small, i.e., $\mid \nabla \boldsymbol{x}(t) \mid \simeq 0$, and also to preserve the discontinuity when $\mid \nabla \boldsymbol{x}(t) \mid$ is strong enough. An example of a ϕ which checks these criteria is $\phi(z) = z$, and $\phi = sqrt1 + z^2$. The choice $\phi(z) = z$ corresponds to the case of total variation which will be adopted in this chapter.

The natural space for dealing with the variational problem (8) is $BV(\Omega)$, the space of all real functions with bounded variation in Ω.

The problem (8) has a unique solution in the space $BV(\Omega)$. The $\tilde{\boldsymbol{x}}$ solution of the variational problem (8) is characterized by the following Euler Lagrange equation

$$\tilde{\boldsymbol{x}}(t) - \bar{\boldsymbol{x}}(t) - \lambda \mathrm{div}\left(\phi'(\mid \nabla \tilde{\boldsymbol{x}}(t) \mid) \frac{\nabla \tilde{\boldsymbol{x}}(t)}{\mid \nabla \tilde{\boldsymbol{x}}(t) \mid} \right) = 0 \tag{9}$$

In the following we will see how to solve this kind of problem numerically, taking $\phi(z) = z$.

3.2 Discretization

The discretization is an essential step to solve the minimization problem (8). In the following, the continuous signals are sampled at a period T_e. We associate each signal $u(t), t \in [0, T]$, with a vector $U := (u(1), \ldots, u(N))^\top \in \cdot E = \mathbb{R}^N$ such that

$$u(k) = u(kT_e), \ \forall k = \{1, \ldots, N\}, \tag{10}$$

The space $\mathbb{E}$ has the scalar product

$$\langle U, V \rangle := \sum_{k=1}^{N} u(k)v(k), \ \forall U, V \in \mathbb{E}. \tag{11}$$

The discrete gradient noted $(\cdot)'$ can be defined as the first derivative of $U \in \mathbb{E}$, noted $U' := (u'(1), \ldots, u'(N))^\top$, where

$$u'(k) = \begin{cases} \dfrac{u(k+1) - x(k)}{Te}, & \text{si } 1 < k < N-1, \\ 0 & \text{si } k = 0. \end{cases} \tag{12}$$

We also define the divergence operator, denoted $(\cdot)^*$, as the adjoint of of the discrete operator $-(\cdot)'$ (the left derivative). For all $U, V \in \mathbb{E}$, we have

$$\langle U^*, V \rangle = -\langle U, V' \rangle. \tag{13}$$

The divergence $U \in \mathbb{E}$, noted $U^* := (u^*(1), \ldots, u^*(N))^\top$, is given by

$$u^*(k) = \begin{cases} \dfrac{u(k) - u(k-1)}{T_e}, & \text{si } 0 < k < N-1, \\ \dfrac{u(1)}{T_e} & \text{si } k = 0, \\ -\dfrac{u(N-1)}{T_e} & \text{si } k = N. \end{cases} \tag{14}$$

We extend all these definitions in a natural way to the elements of the space $\mathbb{E}^p = \mathbb{R}^p \times \mathbb{R}^N$. The elements of the vector space $\mathbb{E}$ will be denoted by capital symbols, while the elements of the vector space $\mathbb{E}$ are denoted by double capital symbols, which can be considered as matrices of dimension $p \times N$.

To solve the problem with TV regularization, several algorithms have been developed. In our study we propose the Chambolle projection algorithm.

3.3 Algorithme de Projection de Chambolle

We consider the following normalized variational problem

$$x_i = \arg \min_{w_i \in BV([0,T])} \left\{ \frac{1}{T} \int_0^T \frac{1}{2} (w_i(t) - \bar{x}_i(t))^2 \, dt + \lambda \frac{1}{T} \int_0^T |w_i'(t)| \, dt \right\}, \; i = 1, \ldots, p. \tag{15}$$

In this paragraph, we show how to determine the observations

$$\mathbb{X} := \left(\boldsymbol{x}(n) = (x_1(n), \ldots, x_p(n))^\top, n = 1, \ldots, N \right), \tag{16}$$

from noisy observations

$$\overline{\mathbb{X}} := \left(\bar{\boldsymbol{x}}(n) = (\bar{x}_1(n), \ldots, \bar{x}_p(n))^\top, n = 1, \ldots, N \right). \tag{17}$$

This will be obtained by solving the discrete version of the problem (15)

$$x_i = \arg\min_{w_i \in \mathbb{E}} \left\{ \frac{1}{N}\sum_{n=1}^{N}\frac{1}{2}(w_i(n) - \bar{x}_i(n))^2 + \lambda\frac{1}{N}\sum_{n=1}^{N}|w_i'(n)| \right\}, \; i = 1, \ldots, p. \quad (18)$$

We start with the following proposition which gives a simple characterization of the solution.

The problem (18) has a unique solution given by

$$\mathbb{X} = \overline{\mathbb{X}} - \Pi_{\lambda G}\left[\overline{\mathbb{X}}\right], \quad (19)$$

where $\Pi_{\lambda G}$ est the orthogonal projection operator on the convex λG, with

$$G := \left\{\mathbb{V}^*/\mathbb{V} \in \mathbb{E}^p, \mid v_i(n) \mid \leq 1 \; \forall (i, n) \in \{1, \ldots, p\} \times \{1, \ldots, N\}\right\}. \quad (20)$$

In order to compute $\mathbb{X}$, we compute the projection operator $\Pi_{\lambda G}$ on the convex set λG, i.e., we compute the projection of the observed signal $\overline{\mathbb{X}}$ on λG which is equivalent to solving the following minimization problem

$$\mathbb{X} = \arg\min_{\mathbb{W} \in \lambda G} \|\mathbb{W} - \overline{\mathbb{X}}\|^2. \quad (21)$$

Otherwise, let D be a convex defined by

$$D := \{V \in \mathbb{E}/\forall k = 1, \ldots, N, |v(k)| \leq 1\}, \quad (22)$$

then the problem (21) is equivalent to

$$\mathbb{X} = \arg\min_{\mathbb{V} \in D^p} \|\lambda \mathbb{V}^* - \overline{\mathbb{X}}\|^2. \quad (23)$$

Let $\rho > 0$, then the optimality condition associated to the problem (23) can be written as follows

$$V_i = \Pi_D\left[V_i - 2\lambda\rho(\lambda V_i^* - \overline{X}_i)'\right], \; \forall i := 1, \ldots, p, \quad (24)$$

where Π_D is the orthogonal projection on D. It is easy to see that for all $\mathbb{V} \in \mathbb{E}$ and for all $k = 1, \ldots, N$,

$$\left(\Pi_D\left[V\right]\right)(k) := \begin{cases} \dfrac{v(k)}{|v(k)|} & \text{si } |v(k)| \geq 1, \\ v(k) & \text{sinon.} \end{cases} \quad (25)$$

In order to solve the problem (23), we will use the classical projection algorithm. For all $i = 1, \ldots, p$, let $X_i^n = 0$. For $n \geq 0$,

$$X_i^{n+1} = \Pi_D \left[X_i^n - 2\lambda\rho(\lambda(X_i^n)^* - \overline{X}_i)' \right], \tag{26}$$

where $\rho > 0$ is well chosen. Then, we have

$$\begin{aligned}
\|X_i^{n+1} - X_i\| &= \|\Pi_D \left[X_i^n + 2\lambda^2\rho((X_i^n)^* - \overline{X}_i)' \right] - \Pi_D \left[X_i + 2\lambda^2\rho((X_i)^* - \overline{X}_i)' \right] \| \\
&\leq \|X_i^n - X_i + 2\lambda^2\rho((X_i^n - X_i)^*)')\| \\
&\leq \| \left[I_d + 2\lambda^2\rho\Delta \right] (X_i^n - X_i)\| \\
&\leq |I_d + 2\lambda^2\rho\Delta| \|X_i^n - X_i\|,
\end{aligned} \tag{27}$$

where Δ is the operator of the second derivative defined by: $\forall V \in \mathbb{E}, \Delta(V) := (V^*)'$. Since the eigenvalues of the operator Δ are negative, then the sequence $\|X_i^n - X_i\|$ is necessarily decreasing. It is therefore convergent if

$$\lambda^2\rho < \frac{1}{\|\Delta\|}, \tag{28}$$

where $\|\Delta\|$ is defined by

$$\|\Delta\| := \sup_{V \in \mathbb{E}} \frac{\langle -\Delta V, V \rangle}{\|V\|^2}. \tag{29}$$

Using (13), we have, $\forall V \in \mathbb{R}^p$,

$$\langle -\Delta V, V \rangle = \|V'\|^2 = \sum_{k-1}^{N-1} \left(\frac{v(k+1) - v(k)}{Te} \right)^2 \leq \frac{4}{Te^2} \|V\|^2. \tag{30}$$

Therefore $\|X_i^n - X_i\|$ converges if:

$$\lambda^2\rho < \frac{T_e^2}{4}. \tag{31}$$

4 Second Step: BSS Based on the Combination of Copula and TV

The objective of this step is to build the estimated sources $\widehat{s}(t) := \widehat{\mathcal{B}} x(t)$, from the partially denoised observations $x = (x_1, \ldots, x_p)^\top$, obtained by solving (8). Our approach consists in minimizing, with respect to $\mathcal{B}$, an estimate of a criterion of the form

$$\mathcal{J}(y) := \mathcal{J}_{\text{sep}}(y) + \mathcal{J}_{\text{reg}}(y), \tag{32}$$

where $y := \mathcal{B}x(t)$, the first term $\mathcal{J}_{\text{sep}}$ is a separation criterion, and $\mathcal{J}_{\text{reg}}$ is the TV regularization term, given by

$$\mathcal{J}_{\text{reg}}(y) := \frac{\gamma}{2T} \int_0^T \|y(t) - \bar{y}(t)\|^2 \, dt + \frac{\nu}{T} \int_0^T |\nabla y(t)| \, dt, \tag{33}$$

where $\gamma > 0$ and $\mu > 0$ are regularization parameters. The term $\mathcal{J}_{\text{sep}}$ is a measure of dependence, or similarity between copula densities. It depends on whether the source components are independent or dependent. In order to clarify this step more precisely, we will study separately the case where the source components are independent, and the case where they are dependent. First, we consider the following stochastic modeling. We assume that the sources and the observations are stationary, so that the sampled versions, with a period T_e,

$$s(n) := s(nT_e),\ x(n) := x(nT_e) \text{ et } y(n) := y(nT_e) = \mathcal{B}x(n),\ n = 1, \ldots, N,$$

can be seen as realizations of random vectors in $\mathbb{R}^p$, noted S, X and $Y := \mathcal{B}X$, respectively.

4.1 Case of Independent Sources

Let's start by defining the separation criterion. We have seen in the previous chapter that if $\mathcal{A}$ is invertible, if the sources are statically independent and if at most one source is Gaussian, a solution to this problem, with an indeterminacy of the scaling factor and a permutation, consists in looking for a matrix $\mathcal{B}$ which makes the components of $Y = \mathcal{B}X$ independent in the statistical sense, by minimizing with respect to $\mathcal{B}$ the following criterion

$$\begin{aligned} \mathcal{J}_{\text{sep}}^{\text{ind}}(Y) :&= \mathbb{KL}\,(c_Y \parallel c_\perp) \\ &= \mathbb{E}\left[\log c_Y(F_{Y_1}(Y_1), \ldots, F_{Y_p}(Y_p)\right]. \end{aligned} \tag{34}$$

where $c^\perp(\cdot)$ is the density of the independence copula and $c_Y(\cdot)$ is the density of the copula of the vector Y.

In order to achieve the separation, the idea is to minimize with respect to $\mathcal{B}$ a statistical estimate $\widehat{\mathcal{J}_{\text{ind}}}(\cdot)$, of $\mathcal{J}_{\text{ind}}(\cdot) = \mathcal{J}_{\text{sep}}^{\text{ind}}(\cdot) + \mathcal{J}_{\text{reg}}(\cdot)$, constructed from the data $y(1), \ldots, y(N)$.

We propose to estimate the criterion $\mathcal{J}_{\text{ind}}(\cdot)$ by

$$\widehat{\mathcal{J}_{\text{ind}}}(y) := \widehat{\mathcal{J}_{\text{sep}}^{\text{ind}}}(y) + \mathcal{J}_{\text{reg,d}}(y), \tag{35}$$

where

$$\mathcal{J}_{\text{reg,d}}(\boldsymbol{y}) := \frac{\gamma}{2N}\sum_{n=1}^{N}(\boldsymbol{y}(n) - \overline{\boldsymbol{y}}(n))^2 + \frac{\mu}{N}\sum_{n=1}^{N}|\nabla \boldsymbol{y}(n)|, \tag{36}$$

is the discrete version of $\mathcal{J}_{\text{reg}}(\cdot)$, and $\widehat{\mathcal{J}_{\text{sep}}^{\text{ind}}}(\cdot)$ is an estimate of the separation criterion $\mathcal{J}_{\text{sep}}^{\text{ind}}(\cdot)$, defined by

$$\widehat{\mathcal{J}_{\text{sep}}^{\text{ind}}}(\boldsymbol{y}) := \frac{1}{N}\sum_{n=1}^{N}\log\left(\widehat{c}_Y(\widehat{F}_{Y_1}(y_1(n)), \ldots, \widehat{F}_{Y_p}(y_p(n)))\right). \tag{37}$$

The estimators of the copula density and the marginal distribution functions are defined as before. The separation matrix is then estimated by

$$\widehat{\boldsymbol{\mathcal{B}}} = \arg\min_{\boldsymbol{\mathcal{B}}}\widehat{\mathcal{J}_{\text{ind}}}(\boldsymbol{y}), \tag{38}$$

which leads to the estimation of the sources

$$\widehat{\boldsymbol{s}}(n) = \widehat{\boldsymbol{\mathcal{B}}}\,\boldsymbol{x}(n), \quad n = 1, \ldots, N.$$

In order to calculate the estimation of the separation matrix $\widehat{\boldsymbol{\mathcal{B}}}$, we use the gradient descent algorithm. This requires the computation of the gradient of the criterion $\widehat{\mathcal{J}_{\text{ind}}}(\cdot)$ with respect to $\boldsymbol{\mathcal{B}}$. The explicit form of the gradient is given by

$$\begin{aligned}\frac{d\widehat{\mathcal{J}_{\text{ind}}}(\boldsymbol{y})}{d\boldsymbol{\mathcal{B}}} :=& \frac{1}{N}\sum_{n=1}^{N}\frac{\frac{d}{d\boldsymbol{\mathcal{B}}}\widehat{\boldsymbol{c}}_Y(\widehat{F}_{Y_1}(y_1(n)), \ldots, \widehat{F}_{Y_p}(y_p(n)))}{\widehat{\boldsymbol{c}}_Y(\widehat{F}_{Y_1}(y_1(n)), \ldots, \widehat{F}_{Y_p}(y_p(n)))}\\ &+ \frac{\gamma}{N}\sum_{n=1}^{N}(\boldsymbol{y}(n) - \overline{\boldsymbol{y}}(n))(\boldsymbol{x}(n) - \overline{\boldsymbol{x}}(n))^\top + \frac{\mu}{N}\sum_{n=1}^{N}\left(\frac{\nabla \boldsymbol{y}(n)}{|\nabla \boldsymbol{y}(n)|}\right)^* \boldsymbol{x}(n)^\top,\end{aligned} \tag{39}$$

where, $\dfrac{d}{d\boldsymbol{\mathcal{B}}} := \dfrac{\partial}{\partial \boldsymbol{\mathcal{B}}_{ij}}$, $i, j = 1, \ldots, p$, et

$$\begin{aligned}\frac{\partial \widehat{c}_{\boldsymbol{Y}}\left(\widehat{F}_{Y_1}(y_1(n)), \ldots, \widehat{F}_{Y_p}(y_p(n))\right)}{\partial \boldsymbol{\mathcal{B}}_{ij}} =& \frac{1}{NH_1\cdots H_p}\sum_{m=1}^{N}\Bigg(\prod_{j=1, j\neq i}^{p} k\left(\frac{\widehat{F}_{Y_j}(y_j(m)) - \widehat{F}_{Y_j}(y_j(n))}{H_j}\right)\\ &\times k'\left(\frac{\widehat{F}_{Y_i}(y_i(m)) - \widehat{F}_{Y_i}(y_i(n))}{H_i}\right)\frac{1}{H_i}\,\frac{\partial(\widehat{F}_{Y_i}(y_i(m)) - \widehat{F}_{Y_i}(y_i(n)))}{\partial \boldsymbol{\mathcal{B}}_{ij}}\Bigg),\end{aligned} \tag{40}$$

with

$$\frac{\partial(\widehat{F}_{Y_i}(y_i(m))}{\partial \boldsymbol{\mathcal{B}}_{ij}} = \frac{1}{Nh_j}\sum_{l=1}^{N}k\left(\frac{y_i(l) - y_i(m)}{h_i}\right)(x_j(l) - x_j(m)), \tag{41}$$

and

$$\frac{\partial(\widehat{F}_{Y_i}(y_i(n))}{\partial \mathcal{B}_{ij}} = \frac{1}{Nh_j}\sum_{l=1}^{N} k\left(\frac{y_i(l) - y_i(n)}{h_i}\right)(x_j(l) - x_j(n)). \tag{42}$$

4.2 Case of Dependent Sources

The sources $\boldsymbol{S}$ are supposed to be dependent by some structure which is represented by a semi-parametric copula model $\mathbb{C}_S \in \{\mathbb{C}(\cdot, \boldsymbol{\theta});\ \boldsymbol{\theta} \in \Theta \subset \mathbb{R}^d\}$. This method has the advantage of being naturally extensible to separate mixtures of dependent sources. Indeed, this approach does not impose any constraint on the independence or the non-gaussianity of the sources. This method is based on the minimization with respect to $\mathcal{B}$ of a non-parametric estimate of the Kulback-Leibler divergence between the non-parametric copula density $\boldsymbol{c}_{\boldsymbol{Y}}(\cdot)$ of the vector $\boldsymbol{Y} = \boldsymbol{B}\boldsymbol{X}$ and a semi-parametric copula density $\boldsymbol{c}(\cdot, \boldsymbol{C})$ which models the dependence structure of the source vector $\boldsymbol{S}$. Our approach is to find the separation matrix $\mathcal{B}$, by minimizing, in relation to $\mathcal{B}$, the estimation of the following criterion

$$\begin{aligned} \mathcal{J}_{\text{sep}}^{\text{dep}}(\boldsymbol{Y}) :=& \mathbb{KL}\,(\boldsymbol{c} \parallel \boldsymbol{c}_{\boldsymbol{Y}}) \\ :=& \mathbb{E}\left[\log\left(\frac{\boldsymbol{c}_{\boldsymbol{Y}}(F_{Y_1}(Y_1), \ldots, F_{Y_p}(Y_p))}{\boldsymbol{c}(F_{Y_1}(Y_1), \ldots, F_{Y_p}(Y_p), \boldsymbol{\theta})}\right)\right]. \end{aligned} \tag{43}$$

The copula $\boldsymbol{c}(\cdot, \boldsymbol{\theta})$ is determined, by applying one of the three procedures presented in [9, 17]. The separation is achieved by minimizing with respect to $\mathcal{B}$ a statistical estimate $\widehat{\mathcal{J}_{\text{dep}}}$, of the criterion $\mathcal{J}_{\text{dep}}(\cdot) = \mathcal{J}_{\text{sep}}^{\text{dep}}(\cdot) + \mathcal{J}_{\text{reg}}(\cdot)$, constructed from the data $\boldsymbol{y}(1), \ldots, \boldsymbol{y}(N)$. We propose to estimate the criterion $\mathcal{J}_{\text{dep}}(\cdot)$ by

$$\widehat{\mathcal{J}_{\text{dep}}}(\boldsymbol{y}) := \widehat{\mathcal{J}_{\text{sep}}^{\text{dep}}}(\boldsymbol{y}) + \mathcal{J}_{\text{reg,d}}(\boldsymbol{y}), \tag{44}$$

where

$$\mathcal{J}_{\text{reg,d}}(\boldsymbol{y}) := \frac{\gamma}{2N}\sum_{n=1}^{N}(\boldsymbol{y}(n) - \overline{\boldsymbol{y}}(n))^2 + \frac{\mu}{N}\sum_{n=1}^{N}|\nabla \boldsymbol{y}(n)|, \tag{45}$$

is the discrete version of $\mathcal{J}_{\text{reg}}(\cdot)$, and $\widehat{\mathcal{J}_{\text{sep}}^{\text{ind}}}(\cdot)$ is an estimate of the separation criterion $\mathcal{J}_{\text{sep}}^{\text{ind}}(\cdot)$, defined by

$$\widehat{\mathcal{J}_{\text{sep}}^{\text{dep}}}(\boldsymbol{y}) := \frac{1}{N}\sum_{n=1}^{N}\log\left(\frac{\widehat{\boldsymbol{c}}_{\boldsymbol{Y}}(\widehat{F}_{Y_1}(y_1(n)), \ldots, \widehat{F}_{Y_p}(y_p(n)))}{\boldsymbol{c}(\widehat{F}_{Y_1}(y_1(n)), \ldots, \widehat{F}_{Y_p}(y_p(n)), \boldsymbol{\theta})}\right), \tag{46}$$

The estimators of the copula density and the marginal distribution functions are defined as before. The separation matrix is then estimated by

$$\widehat{\mathcal{B}} = \arg\min_{\mathcal{B}} \widehat{\mathcal{J}_{\text{dep}}}(\boldsymbol{y}), \tag{47}$$

which leads to the following estimate of the sources

$$\widehat{s}(n) = \widehat{\mathcal{B}}\boldsymbol{x}(n), \quad n = 1, \ldots, N.$$

The solution $\widehat{\mathcal{B}}$, which is the estimation of the separation matrix, can be computed by using the gradient descent algorithm. The expression of the gradient of $\widehat{\mathcal{J}_{\text{dep}}}(\cdot)$ with respect to $\mathcal{B}$ is computed explicitly in the same way as in the previous section

$$\begin{aligned}\frac{d\widehat{\mathcal{J}_{\text{dep}}}(\boldsymbol{y})}{d\mathcal{B}} :=& \frac{1}{N}\sum_{n=1}^{N} \frac{\frac{d}{d\mathcal{B}}\widehat{\boldsymbol{c}}_Y(\widehat{F}_{Y_1}(y_1(n)), \ldots, \widehat{F}_{Y_p}(y_p(n)))}{\widehat{\boldsymbol{c}}_Y(\widehat{F}_{Y_1}(y_1(n)), \ldots, \widehat{F}_{Y_p}(y_p(n)))} \\ &+ \frac{\frac{d}{d\mathcal{B}}\boldsymbol{c}(\widehat{F}_{Y_1}(y_1(n)), \ldots, \widehat{F}_{Y_p}(y_p(n)), \boldsymbol{\theta})}{\boldsymbol{c}(\widehat{F}_{Y_1}(y_1(n)), \ldots, \widehat{F}_{Y_p}(y_p(n)), \boldsymbol{\theta})} \\ &+ \frac{\gamma}{N}\sum_{n=1}^{N}(\boldsymbol{y}(n) - \overline{\boldsymbol{y}}(n))(\boldsymbol{x}(n) - \overline{\boldsymbol{x}}(n))^{\top} + \frac{\mu}{N}\sum_{n=1}^{N}\left(\frac{\nabla\boldsymbol{y}(n)}{|\nabla\boldsymbol{y}(n)|}\right)^{*}\boldsymbol{x}(n)^{\top}.\end{aligned} \tag{48}$$

5 Simulation Results

In this section, we present simulation results illustrating the performance of this method. For simplicity, we will treat the case of two sources-two mixtures ($p = 2$). We will consider various types of signals (independent and dependent), using signals generated from various copulas. The results are compared to those of the method proposed by Pham [18] under the same conditions. We processed instantaneous mixtures of four types of sources (centered and normalized):

1. i.i.d. independent sources with uniform marginals.
2. i.i.d. independent sources with BPSK marginals.
3. i.i.d. sources with uniform marginals, of dependent components generated from the FGM copula with $\widehat{\theta} = 0.8$.
4. i.i.d. sources with BPSK marginals, of dependent components generated from the AMH copula with $\widehat{\theta} = 0.7$.

In all examples, the number of samples is N = 2500. The two sources are mixed with the matrix $\mathcal{A} := [1\ 0.8; 0.8\ 1]$. A centered Gaussian noise with a standard deviation of 0.1 was added to the normalized mixtures, so that the signal to noise ratio is equal to 20 dB. The gradient descent parameter is taken equal to $\nu = 0.1$. In the denoising step, we take the gradient descent parameter is taken equal to $\rho = 0.01$,

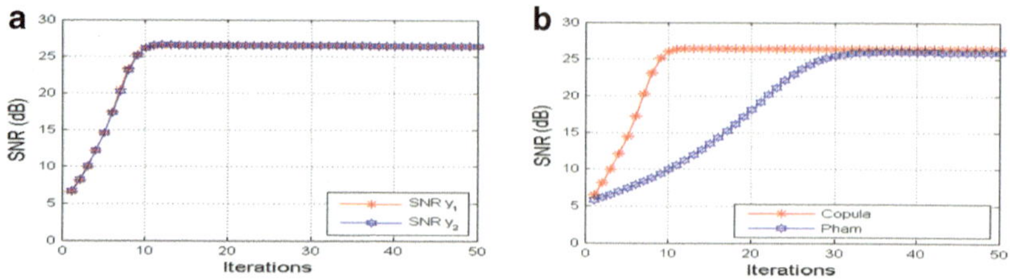

Fig. 1 Separation of sources with uniform marginals, independent components. **a** SNR of the two sources obtained by our approach (independence case), **b** SNR obtained by our approach compared to that of Pham

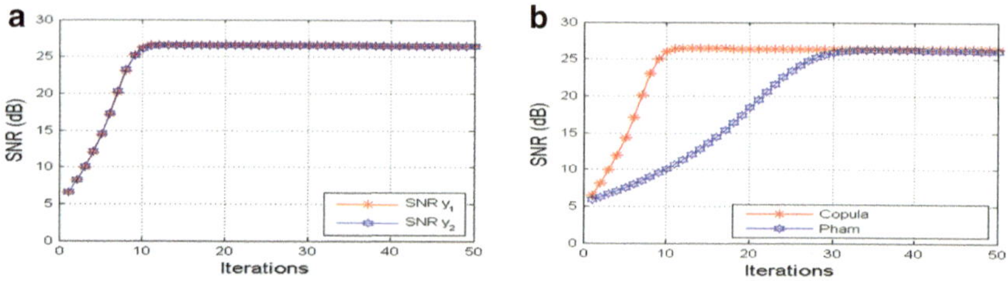

Fig. 2 Separation of the sources with BPSK marginals, of independent components. **a** SNR of the two sources obtained by our approach (independence case), **b** SNR obtained by the approach compared to that of Pham

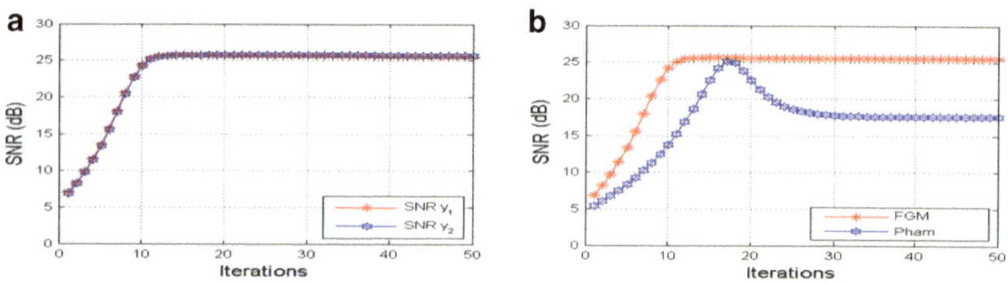

Fig. 3 Separation of dependent sources by the FGM copula, with BPSK marginals. **a** SNR of both sources obtained by our approach for dependent sources, **b** SNR obtained by our approach for dependent sources compared to Pham's

$\lambda = 0.01$, et $\varepsilon = 0.001$. In the second step, we have chosen $\gamma = 0.001$ and $\mu = 0.01$. All simulations are repeated 50 times, and the accuracy of the source estimate is evaluated by the signal-to-residue ratio SNR(dB).

The simulations provided by our approach and by the method proposed by [18] are presented in Figs. 1a, b, 2a, b, 3a, b and 4a, b. In order to facilitate the reading of the results, we have represented in Table 1, the SNR obtained in output relative to the sources S_1 and S_2.

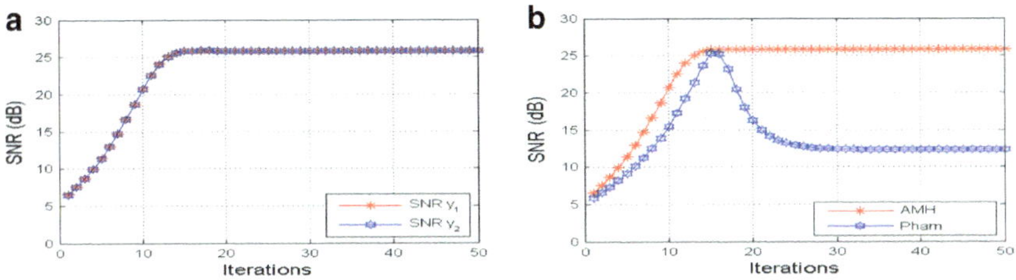

Fig. 4 Separation of dependent sources by AMH copula, with BPSK marginals. **a** SNR of both sources obtained by our approach for dependent sources, **b** SNR obtained by the approach for dependent sources compared to Pham's

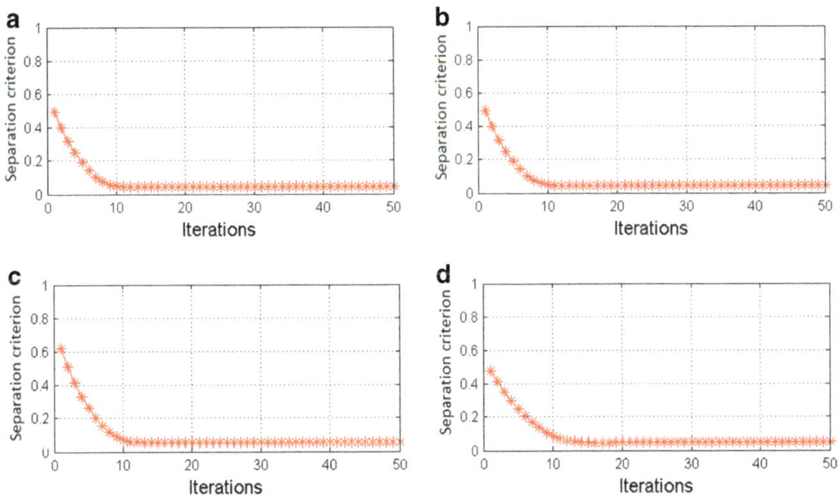

Fig. 5 The separation criterion as a function of the number of iterations. **a** Independence copula (uniform marginals), **b** Independence copula (BPSK marginals), **c** FGM copula (BPSK marginals), **d** Copula AMH (marginal BPSK)

We can see in the standard case where the sources are of independent components, that the SNR of the two methods are almost all higher than 25 dB and that the performances are very good (Figs. 1a, b and 2a, b). On the other hand, in the case where the sources are dependent, we observe from Figs. 3a, b and 4a, b, that our method is able to separate mixtures with a good performance (SNR higher than 25 dB) compared to Pham's method (SNR less than 15 dB). Also, Fig. 5a–d shows that our separation criterion $\widehat{\mathcal{J}_{\text{ind}}}(\cdot)$ converges to 0 when the separation is achieved.

Table 1 SNR obtained in output relative to the sources S_1 and S_2, according to the number of iterations

Types	Kendall's τ	Sources	The approach	Pham's algorithm
Independence (uniform)	0.0002	S_1	26.5170	25.9054
		S_2	26.4779	25.9007
Independence (BPSK)	0.0001	S_1	26.5684	26.0904
		S_2	26.4609	26.1747
FGM (BPSK)	0.1758	S_1	25.6585	19.2595
		S_2	25.7259	15.9124
AMH (BPSK)	0.1935	S_1	25.8491	11.2876
		S_2	25.8773	13.2834

6 Conclusion

In this chapter, we have proposed a new methodology for blind source separation independent/dependent sources for an instantaneous linear mixture in the presence of noise. This approach is based on the minimization of a regularized criterion. The method starts with a denoising step which consists in denoising the observations, using the total variation method. Then, a second step consists in combining the Kullbak-Leibler divergence between the copula densities with the total variation method to separate the mixtures and denoise the estimated sources. From the separation performance point of view, we have shown, through simulations on synthetic data, that our method presents better separation performances.

References

1. P. Comon, C. Jutten, *Handbook of Blind Source Separation: Independent Component Analysis and Applications*. Academic Press (2010)
2. A. Keziou, H. Fenniri, M. Ould Mohamed, G. Delaunay, Séparations aveugle de sources par minimisation des α-divergences, XXIIe colloque GRETSI, Dijon, 8–11 septembre (2009)
3. P. Comon, Independent component analysis, a new concept? Signal Proces. Higher Order Statist. **36**(3), 287–314 (1994)
4. M. Novey, T. Adali, Ica by maximization of nongaussianity using complex functions. Proceedings MLSP
5. A. Belouchrani, A. Cichocki, Robust whitening procedure in blind source separation context. Electron. Lett. **36**(24), 2050–2051 (2000)
6. M. Sahmoudi, H. Snoussi, M. G. Amin, Robust approach for blind source separation in non-gaussian noise environments. Proceedings of ISCCSP, Marrakesh, Morocco, IEEE/EURASIP
7. M. El Rhabi, H. Fenniri, A. Keziou, E. Moreau, A robust algorithm for convolutive blind source separation in presence of noise. Signal Process. **93**(4), 818–827 (2013)

8. A. Ghazdali, M. El Rhabi, H. Fenniri, A. Hakim, A. Keziou, Blind noisy mixture separation for independent/dependent sources through a regularized criterion on copulas. Signal Process. **131**, 502–513 (2017)
9. A. Keziou, H. Fenniri, A. Ghazdali, E. Moreau, New blind source separation method of independent/dependent sources. Signal Process. **104**, 319–324 (2014)
10. N. Mamouni, A. Keziou, H. Fenniri, A. Ghazdali, A. Hakim, A new convolutive source separation approach for independent/dependent source components. Digital Signal Process. **100**, 102701 (2020)
11. A. Ourdou A., Ghazdali, A. Laghrib, A. Metrane, Blind separation of instantaneous mixtures of independent/dependent sources. Circuit. Syst. Signal Proces., 1–24 (2021)
12. A. Ghazdali, A. Hakim, A. Laghrib, N. Mamouni, S. Raghay, A new method for the extraction of fetal ECG from the dependent abdominal signals using blind source separation and adaptive noise cancellation techniques. Theoret. Biol. Med. Modell. **12**(1), 1–20 (2015)
13. A. Ourdou, A. Ghazdali, A. Metrane, M. Hakim, Digital document image restoration using a blind source separation method based on copulas. J. Phys.: Confer. Ser. (Vol. 1743, No. 1, p. 012034). IOP Publishing (2021)
14. A.-N. Tikhonov, V.-Y. Arsenin, *Solutions of Ill-Posed Problems*, V. H. Winston & Sons, Washington, D.C.: Wiley, New Yor (1977)
15. L. Rudin, S. Osher, E. Fatemi, Nonlinear total variation based noise removal algorithms. Physica D: Nonlinear Phenomena **60**(1), 259–268 (1992)
16. A. Chambolle, An algorithm for total variation minimization and applications. J. Math. Imaging Vis. **20**(1–2), 89–97 (2004)
17. A. Ghazdali, M. El Rhabi, H. Fenniri, A. Hakim, A. Keziou, Blind noisy mixture separation for independent/dependent sources through a regularized criterion on copulas. Signal Process. **131**, 502–513 (2017)
18. D. Pham, Mutual information approach to blind separation of stationary sources. IEEE Trans. Inform. Theory **48**(7), 1935–1946 (2002)

Blind Separation of Dependent Sources Using Copula

A. Ourdou, A. Ghazdali, and A. Metrane

Abstract In this chapter, we will be introducing a new efficient Blind Source Separation (BSS) method that handles mixtures of independent and dependent sources. In order to achieve that, we minimize a criterion based on Kullback-Lebiler divergence to set apart the observed mixtures of either dependent or independent sources using copulas as a tool to model this dependency. The proposed algorithm utilizes a gradient descent method to retrieve the signal sources. The efficiency and robustness of this approach is showcased through different experimental results.

Keywords Blind Source Separation · Independent component analysis · Dependent sources · Copulas

1 Introduction

Blind Source Separation (BSS) [5] is the various techniques used to retrieve a set of original sources from their mixed observations with little to no a priori knowledge about the original sources or the mixing environment. In this chapter, we will give a detailed explanation about a new methodology for BSS for both independent and dependent sources of instantaneous linear mixture model, based on the optimization of a criterion between copulas. We introduce three procedures to separate dependent sources according to the degree of the a priori knowledge that we have on the dependency structure. We will show that contrary to the classical BSS methods, we can handle both the case of statistically independent sources and the case of statistically dependent sources (Gaussian or not) as opposed to the most famous approach which is the Independent Component Analysis (ICA) [4, 11, 15] which depends on these two assumptions. Simulation results show the effectiveness of the proposed method on both independent and dependent synthetic data. We have also applied this method to real images and deterministic signals.

A. Ourdou · A. Ghazdali (✉) · A. Metrane
LIPIM, ENSA-Khouribga, Sultan Moulay Slimane University, Beni Mellal, Morocco

A. Nachaoui et al. (eds.), *Mathematical Control and Numerical Applications*,
Springer Proceedings in Mathematics & Statistics 372,
https://doi.org/10.1007/978-3-030-83442-5_11

2 Problem Statement

The model of the BSS problem can be presented as follows: we consider p observations constituted by an instantaneous linear mixture of p sources, in the absence of noise

$$x(t) := As(t) \in \mathbb{R}^p \tag{1}$$

where $s(t) \in \mathbb{R}^p$ is the sources vector; $x(t) \in \mathbb{R}^p$ is the observations vector and $A \in \mathbb{R}^{p\times p}$ is the mixing matrix assumed to be invertible. Blind source separation can be viewed as finding an estimate $\widehat{B}$ of the inverse of A, such that the output vector $\widehat{y} = \widehat{B}x$ is as close as possible to the source vector s.

3 Copula

Understanding the dependence structures between two or more random variables is a widespread practice and copula [10, 14] is the perfect tool to model that dependency. Explicitly, a copula is a function that relates the joint distribution functions $\mathbb{F}$ to the marginal distribution functions F_i, $i = 1, \ldots, p$, as follows:

$$\mathbb{F}(z_1, z_2, \ldots, z_p) := \mathbb{C}(F_1(z_1), F_2(z_2), \ldots, F_p(z_p)),\ z_1, z_2, \ldots, z_p \in \mathbb{R}.$$

Most of the copula theory is based on the famous Sklar theorem [18] which allows to represent any distribution function with a copula. Let $\mathbb{F}_{\mathbf{Z}}(\cdot)$ be a joint distribution function of a random vector $\mathbf{Z}(\cdot)$, and of marginal distribution functions $F_{Z_1}, \ldots, F_{Z_p}$. Then there exists a copula $\mathbb{C}_{\mathbf{Z}}(\cdot)$ such that for all $z \in \mathbb{R}^p$ $\mathbb{F}_{\mathbf{Z}}(z) := \mathbb{C}_{\mathbf{Z}}(F_{Z_1}(z_1), \ldots, F_{Z_p}(z_p))$, if the functions $F_{Z_1}, \ldots, F_{Z_p}$, are continuous, then $\mathbb{C}_{\mathbf{Z}}$ is unique. Conversely, if $\mathbb{C}_{\mathbf{Z}}$ is a copula, and if $F_{Z_1}, \ldots, F_{Z_p}$ are distribution functions, then the function $\mathbb{F}_{\mathbf{Z}}$, is a joint distribution function with the margins $F_{Z_1}, \ldots, F_{Z_p}$.

Sklar's theorem proves that we can associate a copula with each multidimensional distribution function. Let $\mathbb{F}_{\mathbf{Z}}(\cdot)$ be a joint distribution function of a random vector $\mathbf{Z}(\cdot)$, and of marginal distribution functions $F_{Z_1}, \ldots, F_{Z_p}$. Suppose that $F_{Z_1}, \ldots, F_{Z_p}$ are continuous. Then the copula $\mathbb{C}_{\mathbf{Z}}$ associated to $\mathbb{F}_{\mathbf{Z}}$ is given by

$$\mathbb{C}_{\mathbf{Z}}(u_1, \ldots, u_p) := \mathbb{F}_{\mathbf{Z}}(F_{Z_1}^{-1}(u_1), \ldots, F_{Z_p}^{-1}(u_p)), \forall\, (u_1, \ldots, u_p) \tag{2}$$

Among the usual copulas, we have the product copula which has for expression $\mathbb{C}_{\perp}(\boldsymbol{u}) := \prod_{i=1}^{p} u_i$, for any $\boldsymbol{u} = (u_1, \ldots, u_p) \in [0, 1]^p$. Moreover we have, the continuous random variables $Z_1, \ldots, Z_p$ are independent if the dependence structure is the copula product $\mathbb{C}_{\perp}$. It is called the copula of independence. When the copula density exists it is defined as follows:

$$c(\mathbf{u}) = \frac{\partial^p C(\mathbf{u})}{\partial u_1, ..., \partial u_p}, \quad \forall \mathbf{u} \in [0, 1]^p \tag{3}$$

The independence copula density is then equal to 1 on $[0, 1]^p$ and zero otherwise:

$$\boldsymbol{c}^{\perp}(\boldsymbol{u}) := 1_{[0,1]^p}(\boldsymbol{u}), \ \forall \boldsymbol{u} \in [0, 1]^p. \tag{4}$$

Moreover if the probability density function $f_{\mathbf{Y}}$ of the random variable $\mathbf{Y} = (Y_1, \ldots, Y_p)^T$, and their associated marginal probability $f_{Y_1}(.), \ldots, f_{Y_p}(.)$ exist, then we have:

$$f_{\mathbf{Y}}(\mathbf{y}) = \prod_{i=1}^{p} f_{Y_i}(y_i) c_{\mathbf{Y}}(F_{Y_1}(y_1), \ldots, F_{Y_p}(y_p)) \tag{5}$$

Several models for copulas have been suggested in the literature, namely, the semiparametric copula models, which is a wide range class of copula models, that has been used in survival analysis [2, 7]. This class's models rely on parametric copulas that are denoted by $\mathbb{C}(\cdot, \boldsymbol{\theta})$, where $\theta \in \Theta \subset \mathbb{R}^d$, is the indexing parameter, with non-parametric margins. Some of the well known copulas are the copulas of Ali-Mikhail-Haq (AMH) [1], the Frank copula [6] and Clayton copula [3].

4 Case of Independent Sources

In the next two sections, we describe the stated approach that is based on the minimization of a non-parametric estimate of the Kulback-Liebler divergence between the copula densities [12] that was later on extended to the case of noisy mixtures in 2017 [9], the case of convolutive mixtures[13] and in 2021 using the α-divergence instead of the kullback-leibler divergence to better improve the separation results [16]. These copual-based approaches were used for multiple real world applications such as the extraction of fetal ECG from the dependent abdominal signals [8] and digital image restoration [17].

In this section, we will focus on the case where the sources are statistically independent. Let $Y := (Y_1, \cdots, Y_p)^\top \in \mathbb{R}^p$ be a random vector with a joint probability function density f_Y and marginal densities $f_{Y_i}, i = 1, \cdots, p$. Therefore, the mutual information $\mathbb{MI}(Y)$ of the vector Y, could be written as the Kullback-leibler divergence between the copula density of the vector Y, noted $c_Y(.)$ and the copula density of independence $c_\perp$ [12]

$$\begin{aligned}\mathbb{IM}(\boldsymbol{Y}) &= \int_{[0,1]^p} \log\left(\frac{\boldsymbol{c_Y}(\boldsymbol{u})}{1}\right) \boldsymbol{c_Y}(\boldsymbol{u})\, d\boldsymbol{u} =: \mathbb{KL}\left(\boldsymbol{c_Y} \parallel \boldsymbol{c}_\perp\right) \\ &= \mathbb{E}\left[\log \boldsymbol{c_Y}(F_{Y_1}(Y_1), \ldots, F_{Y_p}(Y_p))\right].\end{aligned} \tag{6}$$

From the properties of the Kullback-Leibler divergence, we deduce that the function $B \to \mathbb{KL}(c_Y \| c_\perp)$ is positive, and the minimum 0 is reached in $\widehat{B}$ if and only if the components of Y are independent, i.e., if and only if $\boldsymbol{c}_Y(\boldsymbol{u}) = \boldsymbol{c}_\perp(\boldsymbol{u})$, $\forall \boldsymbol{u} \in [0, 1]^p$.

To achieve the separation, the idea is to minimize a statistical estimate $\widehat{\mathbb{KL}}$ $(\boldsymbol{c}_Y \parallel \boldsymbol{c}_\perp)$, with respect to $\boldsymbol{\mathcal{B}}$, constructed from the data $\boldsymbol{Y}(1), \ldots, \boldsymbol{Y}(N)$. The separation matrix is then estimated by

$$\widehat{\boldsymbol{B}} = \arg\min_{\boldsymbol{\mathcal{B}}} \widehat{\mathbb{KL}}\left(\boldsymbol{c}_Y \parallel \boldsymbol{c}_\perp\right), \tag{7}$$

We propose to estimate the criteria $\mathbb{KL}\left(\boldsymbol{c}_Y \parallel \boldsymbol{c}_\perp\right)$ using

$$\widehat{\mathbb{KL}}\left(\boldsymbol{c}_Y \parallel \boldsymbol{c}_\perp\right) := \frac{1}{N}\sum_{i=1}^{N} \log\left(\widehat{c}_Y(\widehat{F}_{Y_1}(Y_1(n)), \ldots, \widehat{F}_{Y_p}(Y_p(n)))\right), \tag{8}$$

where

$$\widehat{c}_Y(\boldsymbol{u}) := \frac{1}{NH_1 \cdots H_p}\sum_{n=1}^{N}\prod_{i=1}^{p} k\left(\frac{\widehat{F}_{Y_i}(Y_i(n)) - u_j}{H_i}\right), \forall \boldsymbol{u} \in [0, 1]^p, \tag{9}$$

is the kernel estimator of the copula density $c_Y(.)$, and $\widehat{F}_{Y_j}(z)$, $j = 1, \ldots, p$, is the estimate of the marginal distribution function $F_{Y_j}(z)$ of the random variable Y_j in point $z \in \mathbb{R}$, defined by

$$\widehat{F}_{Y_j}(z) := \frac{1}{N}\sum_{n=1}^{N} K\left(\frac{Y_j(n) - z}{h_i}\right), \forall i = 1, \ldots, p, \tag{10}$$

where $K(.)$ is the primitive of the kernel $k(\cdot)$. In our study, the kernel $k(\cdot)$ is chosen to be the standard Gaussian density (zero expectation and unit variance):

$$k(z) := \frac{1}{\sqrt{2\pi}} \exp\left(-\frac{1}{2}z^2\right). \tag{11}$$

where h_i and H_i, $i = 1, \ldots, p$ are called the smoothing windows, and are given by $h_i = \left(\frac{4}{3}\right)^{\frac{1}{5}} N^{\frac{-1}{5}} \widehat{\sigma}_i$ and $H_i = \left(\frac{4}{p+2}\right)^{\frac{1}{p+4}} N^{\frac{-1}{p+4}} \widehat{\Sigma}_i$, where $\widehat{\sigma}_i$ denotes the empirical standard deviation of the sample $Y_i(1), \ldots, Y_i(N)$, and $\widehat{\Sigma}_i$ is the empirical standard deviation of $\widehat{F}_{Y_i}(Y_i(1)), \ldots, \widehat{F}_{Y_i}(Y_i(N))$.

In order to compute $\widehat{\boldsymbol{\mathcal{B}}}$ the estimate of the separation matrix, we use the gradient descent algorithm. This requires the computation of the gradient of the criterion $\widehat{\mathbb{KL}}\left(\boldsymbol{c}_Y \parallel \boldsymbol{c}_\perp\right)$ with respect to $\boldsymbol{\mathcal{B}}$. The explicit form of the gradient is given by

$$\frac{d\widehat{\mathbb{KL}}(\boldsymbol{c_Y} \parallel \boldsymbol{c_\perp})}{d\mathcal{B}} := \frac{1}{N}\sum_{n=1}^{N} \frac{\frac{d}{d\mathcal{B}}\widehat{c}_{\boldsymbol{Y}}\left(\widehat{F}_{Y_1}(Y_1(n)), \ldots, \widehat{F}_{Y_p}(Y_p(n))\right)}{\widehat{c}_{\boldsymbol{Y}}\left(\widehat{F}_{Y_1}(Y_1(n)), \ldots, \widehat{F}_{Y_p}(Y_p(n))\right)}, \tag{12}$$

where $\frac{d}{d\mathcal{B}} := \frac{\partial}{\partial \mathcal{B}_{ij}}$, $i, j = 1, \ldots, p$, and

$$\frac{\partial \widehat{c}_{\boldsymbol{Y}}\left(\widehat{F}_{Y_1}(Y_1(n)), \ldots, \widehat{F}_{Y_p}(Y_p(n))\right)}{\partial \mathcal{B}_{ij}} = \frac{1}{NH_1 \cdots H_p}\sum_{m=1}^{N}\left(\prod_{j=1, j\neq i}^{p} k\left(\frac{\widehat{F}_{Y_j}(Y_j(m)) - \widehat{F}_{Y_j}(Y_j(n))}{H_j}\right)\right.$$
$$\left. \times\, k'\left(\frac{\widehat{F}_{Y_i}(Y_i(m)) - \widehat{F}_{Y_i}(Y_i(n))}{H_i}\right)\frac{1}{H_i}\frac{\partial(\widehat{F}_{Y_i}(Y_i(m)) - \widehat{F}_{Y_i}(Y_i(n)))}{\partial \mathcal{B}_{ij}}\right), \tag{13}$$

with,

$$\frac{\partial(\widehat{F}_{Y_i}(Y_i(m))}{\partial \mathcal{B}_{ij}} = \frac{1}{Nh_j}\sum_{l=1}^{N} k\left(\frac{Y_i(l) - Y_i(m)}{h_i}\right)(X_j(l) - X_j(m)), \tag{14}$$

and

$$\frac{\partial(\widehat{F}_{Y_i}(Y_i(n))}{\partial \mathcal{B}_{ij}} = \frac{1}{Nh_j}\sum_{l=1}^{N} k\left(\frac{Y_i(l) - Y_i(n)}{h_i}\right)(X_j(l) - X_j(n)). \tag{15}$$

We summarize our methodology above, in the following Algorithm 1

Algorithm 1: Separation of independent sources

- ▶ **Input** $\boldsymbol{X}$ observation vector
- ▶ **Output** $\boldsymbol{Y}$ estimated source vector
- ▶ **Begin**
  - • **Initialization:** $\mathcal{B}^0 = \mathcal{I}_d$, $\boldsymbol{Y}^0 = \mathcal{B}^0\boldsymbol{X}$. Given $\varepsilon > 0$ et $\mu > 0$.
  - • **Loop**
    - · $\mathcal{B}^{k+1} = \mathcal{B}^k - \mu \left.\frac{d\widehat{\mathbb{KL}}(\boldsymbol{c_Y} \parallel \boldsymbol{c_\perp})}{d\mathcal{B}}\right|_{\mathcal{B}=\mathcal{B}^k}$
    - · $\boldsymbol{Y}^{k+1} = \mathcal{B}^{k+1}\,\boldsymbol{X}$
  - • **Until** $||\mathcal{B}^{(k+1)} - \mathcal{B}^{(k)}|| < \epsilon$
- ▶ **End**

  $\boldsymbol{Y} = \boldsymbol{Y}^{k+1}$.

5 Case of Dependent Sources

In this case, the $\boldsymbol{S}$ sources are supposed to be dependent by some structure which is represented by an unknown semi-parametric copula $\{\mathbb{C}_{\boldsymbol{S}}(\cdot, \boldsymbol{\theta});\ \boldsymbol{\theta} \in \Theta \subset \mathbb{R}^d\}$. The separation method is based on the minimization with respect to $\boldsymbol{c_Y}$ of a non-parametric estimate of the Kullback-Liebler divergence, between the density of the non-parametric copula $\boldsymbol{c}_{\boldsymbol{Y}}(\cdot)$ of the vector $\boldsymbol{Y} = \boldsymbol{B}\boldsymbol{X}$ and the semi-parametric copula density $\boldsymbol{c}_{\boldsymbol{S}}$ of the source vector $\boldsymbol{S}$:

$$\begin{aligned} \mathbb{KL}\left(\boldsymbol{c}_{\boldsymbol{Y}} \parallel \boldsymbol{c}_{\boldsymbol{S}}\right) :=& \int_{[0,1]^p} \log\left(\frac{\boldsymbol{c}_{\boldsymbol{Y}}(\boldsymbol{u})}{\boldsymbol{c}_{\boldsymbol{S}}(\boldsymbol{u},\boldsymbol{\theta})}\right) \boldsymbol{c}_{\boldsymbol{Y}}(\boldsymbol{u})\, d\boldsymbol{u} \\ =& \mathbb{E}\left[\log\left(\frac{\boldsymbol{c}_{\boldsymbol{Y}}(F_{Y_1}(Y_1),\ldots,F_{Y_p}(Y_p))}{\boldsymbol{c}_{\boldsymbol{S}}(F_{Y_1}(Y_1),\ldots,F_{Y_p}(Y_p),\boldsymbol{\theta})}\right)\right]. \end{aligned} \tag{16}$$

In the following paragraphs, we propose three procedures for mixtures of dependent sources, depending on the availability of source training samples and the knowledge of the source dependency structure.

5.1 Procedure 1: Dependent BSS Based on a Priori Knowledge on the Sources

Suppose that we have a learning sample $\boldsymbol{S}(1),\ldots,\boldsymbol{S}(N) \in \mathbb{R}^p$ from the source vector $\boldsymbol{S}$, distribution $\mathbb{F}_{\boldsymbol{S}}$, marginals $F_{S_1},\ldots,F_{S_p}$, and copula $\mathbb{C}_{\boldsymbol{S}}(\cdot)$. We start by choosing a copula model. To do so, we give ourselves a finite collection of semi-parametric models

$$M_l := \{\boldsymbol{c}_l(.,\boldsymbol{\theta}_l);\ \boldsymbol{\theta}_l \in \Theta_l \subset \mathbb{R}^d\},\ l = 1,\ldots,L, \tag{17}$$

where $\boldsymbol{c}_l$ is a semi-parametric copula density of parameter θ_l. From the learning sample, we choose the most appropriate model among this collection of models in order to estimate the unknown copula $\mathbb{C}_{\boldsymbol{S}}(\cdot)$. We proceed in two steps. First, we compute the semi-parametric maximum likelihood

$$\mathcal{L}_l(\boldsymbol{\theta}_l) := \frac{1}{N}\sum_{n=1}^{N} \log\left[\boldsymbol{c}_l(\widehat{F}_{Y_1}(Y_1(n)),\ldots,\widehat{F}_{Y_p}(Y_p(n)),\boldsymbol{\theta})\right], \tag{18}$$

to determine the parameter

$$\widehat{\boldsymbol{\theta}}_l = \arg\max_{\boldsymbol{\theta}_l} \mathcal{L}_l(\boldsymbol{\theta}_l), \tag{19}$$

for each model M_l, $l = 1,\ldots,L$.

The second step consists in searching for the optimal model M_{l^*} among the collection of models using the AIC criterion, i.e., the model M_{l^*}, with

$$l^* = \arg\min_{l=1,\ldots,L} AIC(M_l), \tag{20}$$

where

$$AIC(M_l) = -2\mathcal{L}_l(\widehat{\boldsymbol{\theta}}_l) + 2k_l, \tag{21}$$

with k_l the number of free parameters of the model. Finally, the copula $\mathbb{C}_S(\cdot)$ will be estimated by the copula $\boldsymbol{C}_{l^*}(\cdot, \widehat{\theta}_{l^*})$.

Let us now turn to the procedure for separating the dependent sources, assuming that the source dependence structure $\mathbb{C}_S(\cdot)$ is presented by the estimated copula $\boldsymbol{c}_{l^*}(\cdot, \widehat{\boldsymbol{\theta}}_{l^*})$. We follow the same procedure for independent sources in the previous section, replacing $\boldsymbol{c}_\perp(\cdot)$ by $\boldsymbol{c}_{l^*}(\cdot, \boldsymbol{l^*})$, we find the separation matrix $\widehat{\mathcal{B}}$, minimizing, with respect to $\mathcal{B}$, a statistical estimate of the following criterion

$$\mathbb{KL}\left(\boldsymbol{c_Y} \parallel \boldsymbol{c_S}\right) = \mathbb{E}\left[\log\left(\frac{\boldsymbol{c_Y}(F_{Y_1}(Y_1), \ldots, F_{Y_p}(Y_p))}{\boldsymbol{c}_{l^*}(F_{Y_1}(Y_1), \ldots, F_{Y_p}(Y_p), \boldsymbol{\theta}_{l^*})}\right)\right]. \tag{22}$$

From the properties of the criterion (22), we deduce that the function $\boldsymbol{B} \mapsto \mathbb{KL}\left(\boldsymbol{c_Y} \parallel \boldsymbol{c_S}\right)$ is positive, and the minimum 0 is reached in $\bar{\mathcal{B}}$ if and only if

$$\boldsymbol{c_Y}(\boldsymbol{u}) = \boldsymbol{c}_{l^*}(\boldsymbol{u}),\ \forall \boldsymbol{u} \in [0, 1]^p.$$

In order to achieve the separation, given the relationship (22), we propose as an estimate $\mathbb{KL}\left(\boldsymbol{c_Y} \parallel \boldsymbol{c_S}\right)$

$$\widehat{\mathbb{KL}}\left(\boldsymbol{c_Y} \parallel \boldsymbol{c_S}\right) := \frac{1}{N}\sum_{n=1}^{N} \log\left(\frac{\widehat{\boldsymbol{c}}_{\boldsymbol{Y}}(\widehat{F}_{Y_1}(Y_1(n)), \ldots, \widehat{F}_{Y_p}(Y_p(n)))}{\boldsymbol{c}_{l^*}(\widehat{F}_{Y_1}(Y_1(n)), \ldots, \widehat{F}_{Y_p}(Y_p(n)), \boldsymbol{\theta}_{l^*})}\right), \tag{23}$$

The estimators of the copula density and the marginal distribution functions are defined as before. The separation matrix is then estimated by

$$\widehat{\mathcal{B}} - \arg\min_{\mathcal{B}} \widehat{\mathbb{KL}}\left(\boldsymbol{c_Y} \parallel \boldsymbol{c_S}\right). \tag{24}$$

The solution $\widehat{\mathcal{B}}$, the estimate of the separation matrix, can be computed by using the gradient descent algorithm. The expression of the gradient of $\widehat{\mathbb{KL}}\left(\boldsymbol{c_Y} \parallel \boldsymbol{c_S}\right)$ with respect to $\mathcal{B}$ is computed in an explicit way in the same way as in the previous section

$$\begin{aligned}\frac{d\widehat{\mathbb{KL}}\left(\boldsymbol{c_Y} \parallel \boldsymbol{c_S}\right)}{d\mathcal{B}} :=& \frac{1}{N}\sum_{n=1}^{N} \frac{\frac{d}{d\mathcal{B}}\widehat{\boldsymbol{c}}_{\boldsymbol{Y}}(\widehat{F}_{Y_1}(Y_1(n)), \ldots, \widehat{F}_{Y_p}(Y_p(n)))}{\widehat{\boldsymbol{c}}_{\boldsymbol{Y}}(\widehat{F}_{Y_1}(Y_1(n)), \ldots, \widehat{F}_{Y_p}(Y_p(n)))} \\ &+ \frac{\frac{d}{d\mathcal{B}}\boldsymbol{c}_{l^*}(\widehat{F}_{Y_1}(Y_1(n)), \ldots, \widehat{F}_{Y_p}(Y_p(n)), \boldsymbol{\theta}_{l^*})}{\boldsymbol{c}_{l^*}(\widehat{F}_{Y_1}(Y_1(n)), \ldots, \widehat{F}_{Y_p}(Y_p(n)), \boldsymbol{\theta}_{l^*})}.\end{aligned} \tag{25}$$

We obtain the following Algorithm 2

Algorithm 2: Separation of dependent sources; The copula of the sources and its parameter are known or estimated.

► **Input**

X observation vector, $c_{l^*}(\cdot, \theta_{l^*})$ copula density of the estimated sources.

► **Output**

Y estimated sources vector

► **Begin**

• **Initialisation:** $\mathcal{B}^0 = \mathcal{I}_d$, $Y^0 = \mathcal{B}^0 X$. Given $\varepsilon > 0$ et $\mu > 0$.

• **Loop**

$$\cdot\ \mathcal{B}^{k+1} = \mathcal{B}^k - \mu \left.\frac{d\widehat{\mathbb{KL}}(c_Y \parallel c_S)}{d\mathcal{B}}\right|_{\mathcal{B}=\mathcal{B}^k}$$

$$\cdot\ Y^{k+1} = \mathcal{B}^{k+1}\, X$$

• **Until** $||\mathcal{B}^{(k+1)} - \mathcal{B}^{(k)}|| < \epsilon$

► **End**

$Y = Y^{k+1}$.

5.2 *Procedure 2: BSS for Dependent Sources When the Dependency Model Is Known and the Parameter Is Unknown*

In this paragraph, we will consider the case where we have some a priori information about the source dependence structure S, represented by a semi-parametric copula model $\mathbb{C}_S(\cdot)$, with an unknown parameter θ. We propose to estimate the separation matrix, by minimizing, with respect to $\mathcal{B}$ and θ, a statistical estimate of the following criterion

$$\begin{aligned}(\mathcal{B}, \theta) \mapsto \mathbb{KL}(c_Y \parallel c_\theta) &:= \int_{[0,1]^p} \log\left(\frac{c_Y(u)}{c_\theta(u)}\right) c_Y(u)\, du \\ &= \mathbb{E}\left[\log\left(\frac{c_Y(F_{Y_1}(Y_1), \dots, F_{Y_p}(Y_p))}{c_\theta(F_{Y_1}(Y_1), \dots, F_{Y_p}(Y_p))}\right)\right].\end{aligned} \tag{26}$$

$\mathcal{B}$ can be estimated, by minimizing, with respect to $\mathcal{B}$ and θ, the estimate of $\mathbb{KL}(c_Y \parallel c_\theta)$ as follows

$$\widehat{\mathbb{KL}}(c_Y \parallel c_\theta) := \frac{1}{N}\sum_{n=1}^{N} \log\left(\frac{\widehat{c}_Y(\widehat{F}_{Y_1}(Y_1(n)), \dots, \widehat{F}_{Y_p}(Y_p(n)))}{c_\theta(\widehat{F}_{Y_1}(Y_1(n)), \dots, \widehat{F}_{Y_p}(Y_p(n)))}\right). \tag{27}$$

Therefore, the estimate of the separation matrix is given by

$$\widehat{B} = \arg\min_{\mathcal{B}} \widehat{\mathbb{KL}}\left(c_Y \parallel c_\theta\right). \tag{28}$$

The solution $\widehat{\mathcal{B}}$ is computed by the gradient descent algorithm. The expression of the gradient of $\widehat{\mathbb{KL}}\left(c_Y \parallel c_\theta\right)$ with respect to $\mathcal{B}$ and θ is computed in an explicit way in the same way as in the previous sections.
We summarize our methodology of procedure 2, in the Algorithm 3 below

Algorithm 3: Separation of dependent sources : the copula is known and the parameter is unknown

► **Input**

X observation vector, copula density model: $\{c_\theta(\cdot);\ \theta \in \Theta\}$

► **Output**

Y estimated sources vector

► **Begin**

• **Initialisation:** $\mathcal{B}^0 = \mathcal{I}_d$, $Y^0 = \mathcal{B}^0 X$. Given θ^0, $\varepsilon > 0$ et $\mu > 0$.

• **Loop**

$$\cdot\ \theta^{k+1} = \theta^k - \mu \left.\frac{d\widehat{\mathbb{KL}}\left(c_Y \parallel c_{\theta^k}\right)}{d\theta}\right|_{\theta=\theta^k}$$

$$\cdot\ \mathcal{B}^{k+1} = \mathcal{B}^k - \mu \left.\frac{d\widehat{\mathbb{KL}}\left(c_Y \parallel c_{\theta^{k+1}}\right)}{d\mathcal{B}}\right|_{\mathcal{B}=\mathcal{B}^k}$$

$$\cdot\ Y^{k+1} = \mathcal{B}^{k+1}\ X$$

• **Until** $||\mathcal{B}^{(k+1)} - \mathcal{B}^{(k)}|| < \epsilon$

► **End**

$Y = Y^{k+1}$.

5.3 *Procedure 3: BSS for Dependent Sources When the Dependency Model Is Unknown and the Parameter Is Unknown*

In this paragraph, we have no information neither on the sources S, nor on the dependency structure $\{\mathbb{C}_S(\cdot, S) \in \Theta \subset \mathbb{R}^d\}$. In order to achieve the separation, we propose to use the following methodology.

First, we consider a class of L models representing the copula density $c_S(\cdot)$ of sources noted

$$M_l := \{c_l(\cdot, \theta_l);\ \theta_l \in \Theta_l \subset \mathbb{R}^d\},\ l = 1, \ldots, L. \tag{29}$$

Each model is defined by a semi-parametric copula density $c_l(\cdot, c_l)$ of parameter c_l. Our goal is to separate the sources using an appropriate model $M_{l^*} = \{c_l(\cdot, \theta_l);\ \theta_l \in \Theta_l \subset \mathbb{R}^d\}$. Let us denote $c_l(\cdot) := c_l(\cdot, c_l)$, and let us assume that each model $M_l, l = 1, \ldots, L,$. Therefore, we propose to apply procedure 2 described in paragraph 5.2 above for each model, and then choose the solution that minimizes the criterion estimate over all models considered, i.e.,

$$\widehat{\mathcal{B}} = \arg\min_{\mathcal{B}} \min_{\theta_{l^*} \in \Theta_{l^*}} \widehat{\mathbb{KL}} (\boldsymbol{c}_Y \parallel \boldsymbol{c}_{l^*}) ,$$

with

$$l^* = \arg \min_{l=1,\ldots,L} \min_{\mathcal{B}} \min_{\theta_l \in \Theta_l} \widehat{\mathbb{KL}} (\boldsymbol{c}_Y \parallel \boldsymbol{c}_l) . \tag{30}$$

The sources are thus estimated by

$$\widehat{\boldsymbol{s}}(t) = \widehat{\mathcal{B}} \boldsymbol{x}(t).$$

6 Experimental Performance Study

We will present experimental results to illustrate the performance of the algorithms we proposed during the theoretical development of the previous sections. We will treat three series of examples of simulation results in the case of two sources-two mixtures ($p = 2$). We will first see in the Sect. 6.1, the results of our algorithms applied to various types of signals (independent and dependent). We will then treat in the series Sect. 6.2 two examples of two types of dependent images.

6.1 *Series 1: Tests with Various Types of Signals*

This first set of tests has been performed using signals generated from various copulas. We concede two classes of sources. The first class of source corresponds to a bivalent phase shift keying called BPSK. The second class is a set of sources with uniform marginals. We have treated instantaneous mixtures of 6 types of sources (centered and normalized):

1. i.i.d. sources with uniform marginals, independent components (Fig. 1a).
2. i.i.d. sources with BPSK marginals, independent components (Fig. 1b).
3. i.i.d. sources with uniform marginals, of dependent components generated from the FGM copula with $\widehat{\theta} = 0.9$ (Fig. 2a).
4. i.i.d. sources with BPSK marginals, dependent components generated from the FGM copula with $\widehat{\theta} = 0.9$ (Fig. 2b).
5. i.i.d. sources with uniform marginals, of dependent components generated from the Clayton copula with $\widehat{\theta} = 1.5$ (Fig. 3a).
6. i.i.d. sources with BPSK marginals, dependent components generated from the Clayton copula with $\widehat{\theta} = 1.5$ (Fig. 3b).

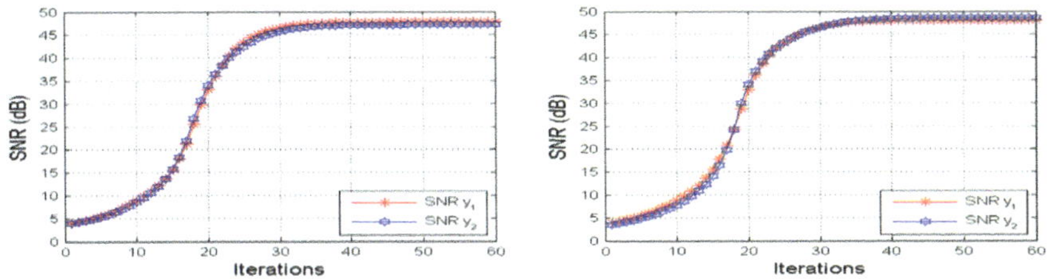

(a) Uniform marginals (b) BPSK marginals

Fig. 1 SNR of the two sources of independent components obtained by the Algorithm 1, in function of the number of iterations

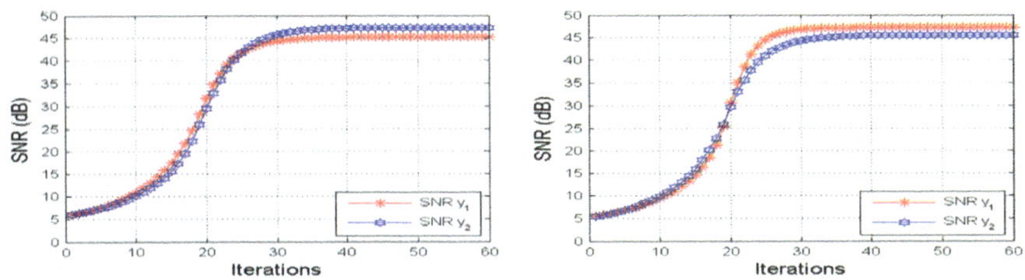

(a) Uniform marginals (b) BPSK marginals

Fig. 2 SNR of the two sources by the FGM copula obtained by the Algorithm 2 (case procedure 1), in function of the number of iterations

The number of samples is fixed at $N = 2500$. We used the following mixing matrix $\mathcal{A} = [1, 0.8; 0.8, 1]$. In order to perform the separation, we used the Algorithm 1 to separate the independent sources, and the first procedure of the Sect. 5 (the copulas and their parameters are known) to separate the dependent sources. The separation quality is evaluated by the SNR in dB, which is the most widely used performance measure in source separation.

We can see in the standard case where the sources are of independent components and the not-standard case, that the SNRs are almost all above 40 dB and that the performances are good (Figs. 1a, b–2a, 3b). Also, Fig. 5 shows that our separation criterion $\widehat{\mathbb{KL}}(\cdot \parallel \cdot)$ converges to 0 when the separation is achieved (Fig. 4 and Table 1).

6.2 Series 2: Tests with Images

This second test presents the results obtained with a mixture of two images of size 512×512, "Lena" and "Barbara". We consider each image as a source vector of dimension $N_t = 512 \times 512$ pixels. We applied the same mixing matrix as before. The two source images were linearly mixed using $\mathcal{A}$, to form two synthetic observations.

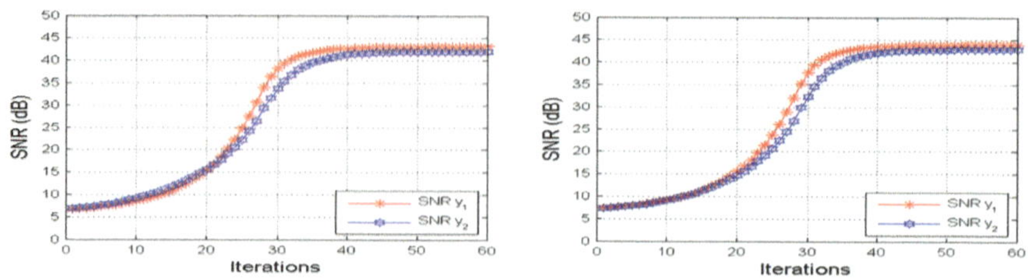

(a) Uniform marginals (b) BPSK marginals

Fig. 3 SNR of the two sources by the Clayton copula obtained by the Algorithm 2 (case procedure 1), in function of the number of iterations

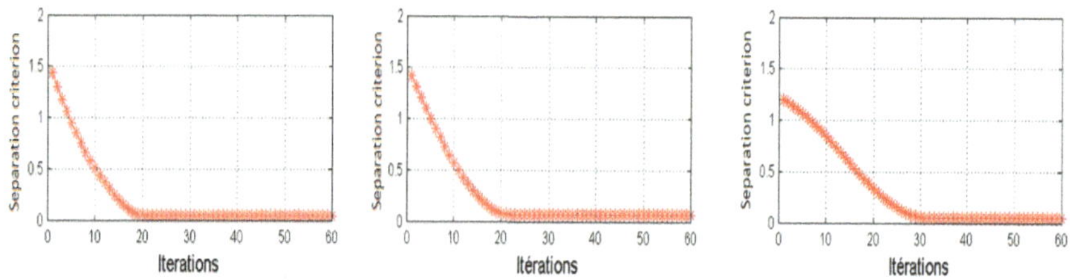

Independence copula FGM Copula Clayton Copula

Fig. 4 The separation criterion value according to the number of iterations

Source 1 Source 2 Observation 1 Observation 2

Fig. 5 The two sources "Lena" and "Barbara", with the observed mixtures of the two images

Table 1 SNR Output between the estimated sources and the original sources

Types	S1	S2
Independent (uniform)	47.7318	47.1218
Independent (BPSK)	47.9826	48.5440
FGM (uniform)	45.1836	47.2113
FGM (BPSK)	47.1603	45.5482
Clayton (uniform)	43.3094	42.1874
Clayton (BPSK)	43.9177	42.8756

Source 1 estimated by la procedure 1.

Source 2 estimated by procedure 1.

Fig. 6 The "Lena" and "Barbara" images estimated by the Sect. 2 algorithm of procedure 1

In order to perform the separation we applied our methodology described in Sect. 5. In order to reduce the computation time, we will take random samples of $N = 3000$ points.

Procedure 1

In this paragraph, we illustrate the performance of the separation method described in procedure 5.1. Suppose that we have a sample of sources to build copula models. Then, we select a semi-parametric model of the copula density $\{\boldsymbol{c}_{\boldsymbol{S}}(F_{S_1}(s_1), F_{S_2}(s_2), \boldsymbol{\theta});\ \boldsymbol{\theta} \in \Theta \subset \mathbb{R}^d\}$ of the $\boldsymbol{S} := (S_1, S_2)^\top$ sources, where S_1 and S_2 represents the random samples of $N = 3000$ points from the first and the second image sources. Applying the model selection step to the three candidate models: FGM, AMH, Clayton, leads to the choice of the AMH copula with $\theta = -0.7697$. Finally we separate the sources by applying our separation criterion described previously by Eq. (25).

We present in Fig. 6 the separation results on our reference mixtures. We can see that the proposed method is able to separate the mixtures with a good performance of an SNR around 40 dB.

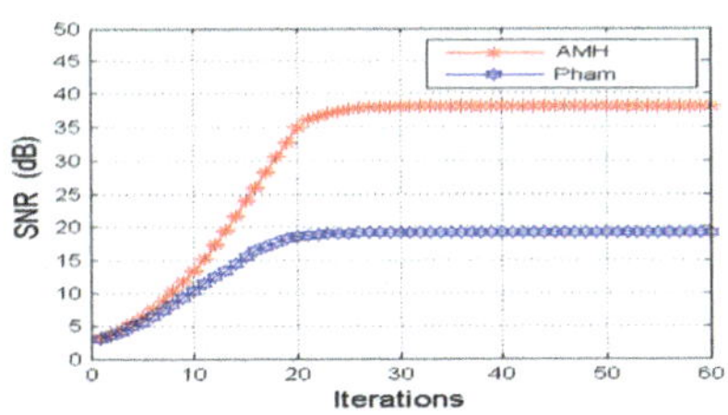

(a) SNR according to the number of iterations.

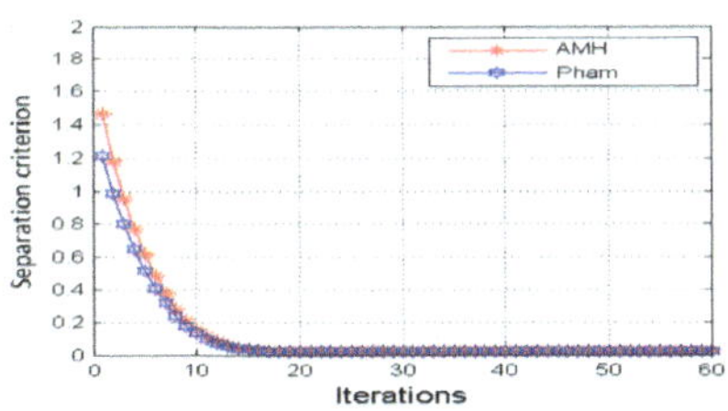

(b) The separation criterion of procedure 1.

Fig. 7 The average SNR and the separation criterion according to the number of iterations using procedure 1

Fig. 8 The images "Lena" and "Barbara" estimated by the Algorithm 3 of procedure 2

Source 1 estimated by procedure 2.

Source 2 estimated by procedure 2.

Also, Fig. 7b shows that our separation criterion $\widehat{\mathbb{KL}}(\cdot \parallel \cdot)$ converges to 0 when the separation is achieved.

Procedure 2

This paragraph illustrates the performance of the separation method described in procedure 5.2. Let us suppose that we have some a priori information on the dependence structure of the sources $\boldsymbol{S}$, represented by the semi-parametric copula $\mathbb{C}_S(\cdot)$, with an unknown parameter $\boldsymbol{\theta}$. In our case the dependency structure of the image sources "Lena" and "Barbara" is supposed to be modeled by the AMH copula with an unknown $\boldsymbol{\theta}$ parameter. Following the procedure 2, we find that the parameter $\boldsymbol{\theta}$ of the AMH copula converges to $\widehat{\theta} = -0.7529$. The separation is done, using the separation criterion described previously by the Eq. (27).

The separation results are shown in Fig. 8. We observe that the proposed method is able to separate the mixtures with a good performance of an SNR around the value 40 dB.

Also, Fig. 9a shows that our separation criterion $\widehat{\mathbb{KL}}(\cdot \parallel \cdot)$ converges to 0 when the separation is achieved. Finally, Fig. 9b presents the convergence of θ as a function of the number of iterations.

Procedure 3

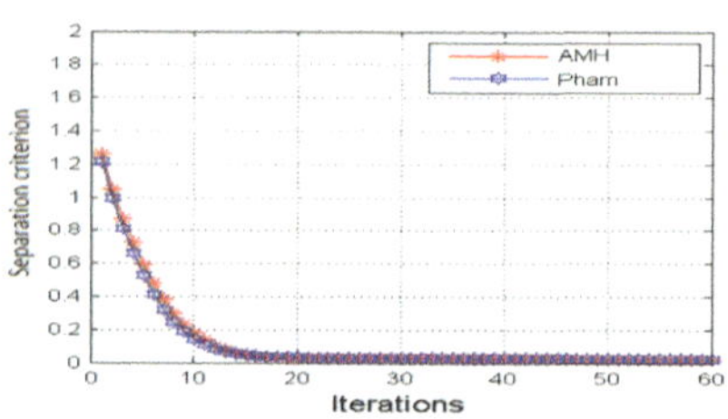

(a) The separation criterion of procedure 2

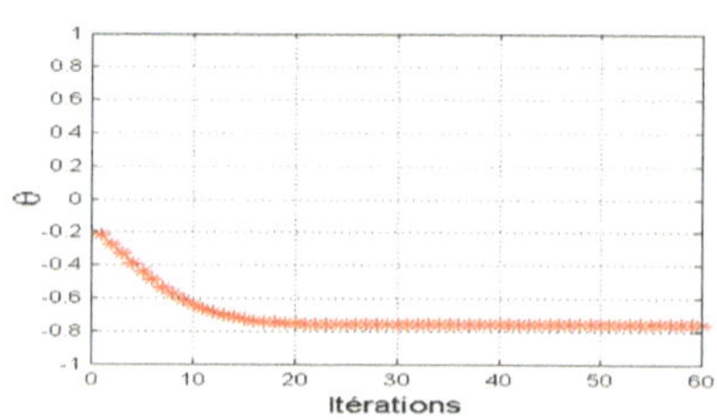

(b) The θ parameter of procedure 2.

Fig. 9 The separation criterion and the value of theta according to the number of iterations using procedure 2

Fig. 10 The "Lena" and "Barbara" images estimated by procedure 3

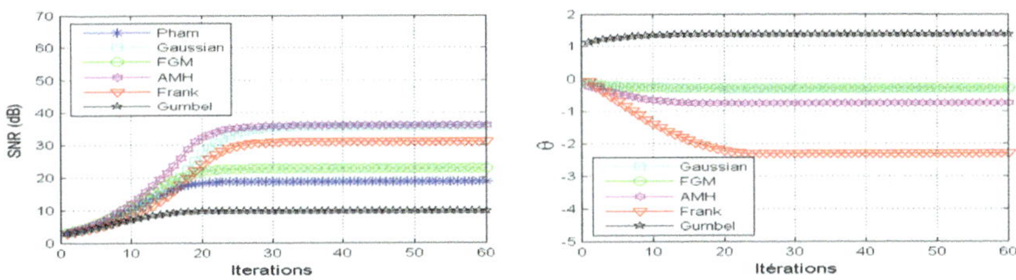

(a) The average SNR of procedure 3. (b) The parameter θ of the procedure 3.

Fig. 11 The average SNR and the value of theta according to the number of iterations using procedure 3

To illustrate the performance of the separation method described in procedure 3, we take the same sources above. Let us suppose that we have no information neither on the sources S, nor on the dependency structure $\mathbb{C}_S(\cdot)$. In order to perform the separation, we apply procedure 3, on the three candidate models defined previously. The optimal model for "Lena" and "Barbara" is given by the AMH copula, with $\widehat{\theta} = -0.7529$.

We illustrate in Fig. 10 the separation results. For more evaluation, the SNR obtained in output relative to the sources S_1 and S_2. We can see that our method gives the right optimal model and good results compared to other models.

Furthermore, Fig. 11a shows The average SNR of procedure 3 for all the copula models. In Fig. 11b, we present the convergence of θ values as a function of the number of iterations.

7 Conclusion

We have described in this chapter a new blind source separation method applied to linear instantaneous mixtures of independent and dependent sources. This method estimates the separation matrix and the sources, based on the optimization of a criterion between copula densities. It does not require either the independence of the sources or their non-gaussianity. Separation results on synthetic data and on real data show the efficiency of the proposed method.

References

1. M. M. Ali, N. N. Mikhail, M. S. Haq, A class of bivariate distributions including the bivariate logistic, J. Multivariate Anal. 8 (3) (1978) 405–412
2. X. Chen, Y. Fan, Estimation and model selection of semiparametric copula-based multivariate dynamic models under copula misspecification, J. Econometrics 135 (1–2) (2006) 125–154
3. D. G. Clayton, A model for association in bivariate life tables and its application in epidemiological studies of familial tendency in chronic disease incidence, Biometrika 65 (1) (1978) 141–151
4. P. Comon, Independent component analysis, a new concept? Signal Process. **36**(3), 287–314 (1994). Higher Order Statistics
5. P. Comon, C. Jutten, *Handbook of Blind Source Separation: Independent Component Analysis and Applications* (Academic press, 2010)
6. M. J. Frank, On the simultaneous associativity of $F(x, y)$ and $x + y - F(x, y)$, Aequationes Math. 19 (2–3) (1979) 194–226
7. C. Genest, K. Ghoudi, L.-P. Rivest, A semiparametric estimation procedure of dependence parameters in multivariate families of distributions, Biometrika 82 (3) (1995) 543–552
8. A. Ghazdali, A. Hakim, A. Laghrib, N. Mamouni, S. Raghay, A new method for the extraction of fetal ECG from the dependent abdominal signals using blind source separation and adaptive noise cancellation techniques. (2015) Theoretical Biology and Medical Modelling, 12(1), 1–20
9. A. Ghazdali, M. El Rhabi, H. Fenniri, A. Hakim, A. Keziou, Blind noisy mixture separation for independent/dependent sources through a regularized criterion on copulas. Signal Process. **131**, 502–513 (2017)
10. H. Joe, *Multivariate Models and Dependence Concepts*, Monographs on Statistics and Applied Probability, vol. 73. (Chapman & Hall, London, 1997)
11. A. Keziou, H. Fenniri, M. Ould Mohamed, G. Delaunay, Séparations aveugle de sources par minimisation des α-divergences, XXIIe colloque GRETSI, Dijon, 8–11 septembre (2009)
12. A. Keziou, H. Fenniri, A. Ghazdali, E. Moreau, New blind source separation method of independent/dependent sources, Signal Processing 104 (2014) 319–324
13. N. Mamouni, A. Keziou, H. Fenniri, A. Ghazdali, A. Hakim, A new convolutive source separation approach for independent/dependent source components. Digital Signal Process. **100**, 102701 (2020)
14. R. B. Nelsen, An introduction to copulas, 2nd Edition, Springer Series in Statistics, Springer, New York, 2006
15. M. Novey, T. Adali, Ica by maximization of nongaussianity using complex functions, Proceedings MLSP
16. A. Ourdou A., Ghazdali, A. Laghrib, A. Metrane, Blind separation of instantaneous mixtures of independent/dependent sources. Circuit. Syst. Signal Process., 1–24 (2021)
17. A. Ourdou, A. Ghazdali, A. Metrane, M. Hakim, Digital document image restoration using a blind source separation method based on copulas. J. Phys.: Confere. Ser. **1743**(1), 012034) (2021). IOP Publishing
18. A. Sklar, Fonctions de répartition à n dimensions et leurs marges, Publ. Inst. Statist. Univ. Paris 8 (1959) 229–231

Printed by Printforce, the Netherlands